Unity 5 for Android Essentials

A fast-paced guide to building impressive games and applications for Android devices with Unity 5

Valera Cogut

PUBLISHING

BIRMINGHAM - MUMBAI

Unity 5 for Android Essentials

First published: August 2015

Production reference: 1310715

Published by Packt Publishing Ltd.
Livery Place
35 Livery Street
Birmingham B3 2PB, UK.

ISBN 978-1-78439-919-1

www.packtpub.com

Credits

Author
Valera Cogut

Reviewers
Todd Bello

Robbyn Blumenschein

Ali Raza

Joe Whitehouse

Commissioning Editor
Amarabha Banerjee

Acquisition Editor
Vinay Argekar

Content Development Editor
Samantha Gonsalves

Technical Editor
Faisal Siddiqui

Copy Editor
Charlotte Carneiro

Project Coordinator
Kinjal Bari

Proofreader
Safis Editing

Indexer
Tejal Soni

Production Coordinator
Aparna Bhagat

Cover Work
Aparna Bhagat

About the Author

Valera Cogut is an independent professional software and video game developer who has worked in the game industry since 2008 and has more than 12 years of programming experience. He is a passionate software and game developer with different areas of expertise. Before diving into the game industry 6 years ago, Valera created websites and applications with PHP, the Yii framework, the Zend framework, Relational Database Management Systems, Apache and Nginx, C#, C++, C, Objective-C, Java, Python, UML, and many other technologies. Having a mathematical background (analytical geometry, linear algebra, logic, statistics and probability, differential equations, graph theory, and mathematical analysis), he finally realized that game development was his mission. Reusable designs, optimized algorithms, clean code, and elaborated workflows—these things make him happy.

Valera has had the opportunity to produce titles for multiple platforms, including Windows Phone, Android, iOS, PC, and Mac. Today, he continues to produce fun and original games, participate in game jams, and author books. He was a technical reviewer for *Unity Android Game Development by Example Beginner's Guide*, *Packt Publishing*.

Currently, Valera has a contract as a Unity3D C# developer for Kaufcom GmbH, a well-known games and apps producer from Switzerland `http://www.kauf.com` (Android games: `https://play.google.com/store/search?q=kaufcom`).

First of all, thanks to Vinay Argekar, Melita Lobo, Samantha Gonsalves, and the Packt Publishing team for creating this book plus many others, thereby helping the Unity community grow and mature. A special thanks goes to my family for the love, education, and freedom they gave me. I would also like to thank my best friends, Paul Kofmann, Alexander Kofmann, and Vitaly Shapovalov. Thanks to the Kaufcom company. Of course, no progress of mine could have been achieved without the support and love of my wife, Irina.

About the Reviewers

Todd Bello is a software engineer at Microsoft working on 3D Maps for Bing and Windows. He graduated from Rochester Institute of Technology in 2013 with a BS in game design and development. He has reviewed a few MSDN articles and the free e-book *Machine Learning Using C# Succinctly* by James McCaffery.

I'd like to thank my friend Leigh Raze.

Robbyn Blumenschein comes from a humble background. She was raised in a military family. Then, she joined the military, but she has always been interested in creating games. She worked for Cryptic Studios, Electronic Arts, I-Play, Kunomi, Aeria Games, and finally, started her own business, 2 Bee Soft LLC.

I always think about my mentor and teacher, Patrick Downey, who passed away in 2007. I will never give up on my dream.

Ali Raza has over 8 years of international industry experience working with companies from US, UK, Australia, Norway, and Brazil in almost every major field of design and development. He has developed applications and games for Windows, Mac, iOS, Android, BlackBerry, and web platforms. He has been a part of social networks, healthcare IT, gaming, industrial automation, business and finance, media, and e-learning, which have enabled him to grow into a versatile developer with a diverse skill set.

He is currently working as a UI/UX program development lead architect in a US-based healthcare IT company, which specializes in cross-platform IT healthcare products.

In addition to that, Ali is also the founder of a start-up company, where he works with the aid of a self-trained team on various cross-platform applications for international clientele in Norway, Brazil, and Italy.

Ali is an Adobe Training Partner and Microsoft Certified Trainer in enterprise, desktop, and web development. He has held several nationwide exclusive official and custom-based training camps on core design methodologies and programming languages at companies, universities, and government organizations.

Ali is a published author and has contributed to key international magazines, books, and examination aides in a wide area of IT topics, including gaming, data visualization, and web technologies. He has authored dozens of articles for Flash & Flex Developer Magazines, Software Development Journal, and Packt Publishing.

Furthermore, Ali also technically reviews IT-based books and publications from time to time.

You can reach him at manofspirit@gmail.com.

I would like to thank Melita Lobo and Kinjal Bari from Packt Publishing for giving me the opportunity to be part of the technical reviewing team. I am also indebted to my long time friend, Sadia Suhail, who helped me with the reviewing process.

Thank you! God bless you all!

Joe Whitehouse has been playing video games since he was a very young child. Growing up, he became more and more fascinated with how they worked or could be broken down. For years, he enjoyed searching for exploits and was one of those to discover the Halo Super Bounce. Many years later, this curiosity and interest in breaking games led to triOS College's video game design and development course. Here, he was taught by Christopher Coey and graduated with distinction. During his time at triOS College, Joe interned at a gaming studio, Little Guy Games, working as a programmer but took on more of a role as a technical tester, finding and fixing bugs. He has also created many video game prototypes using XNA, Unity 2D and 3D, Android, and HTML5 Games.

More recently, he went back to triOS college for their video game design technologies, a course taught by Rocco Commisso. Joe is now learning to create all the visuals in video games so that he can reach another goal: to create more games outside of school with a small team.

I would like to thank Christopher Coey for teaching the programming course and putting up with all my many questions at the start. I would also like to thank Rocco Commisso for everything he taught me as I continue to learn.

www.PacktPub.com

Support files, eBooks, discount offers, and more

For support files and downloads related to your book, please visit www.PacktPub.com.

Did you know that Packt offers eBook versions of every book published, with PDF and ePub files available? You can upgrade to the eBook version at www.PacktPub.com and as a print book customer, you are entitled to a discount on the eBook copy. Get in touch with us at service@packtpub.com for more details.

At www.PacktPub.com, you can also read a collection of free technical articles, sign up for a range of free newsletters and receive exclusive discounts and offers on Packt books and eBooks.

https://www2.packtpub.com/books/subscription/packtlib

Do you need instant solutions to your IT questions? PacktLib is Packt's online digital book library. Here, you can search, access, and read Packt's entire library of books.

Why subscribe?

- Fully searchable across every book published by Packt
- Copy and paste, print, and bookmark content
- On demand and accessible via a web browser

Free access for Packt account holders

If you have an account with Packt at www.PacktPub.com, you can use this to access PacktLib today and view 9 entirely free books. Simply use your login credentials for immediate access.

Table of Contents

Preface

Learn the Unity 5 engine to master designing and building awesome real-world games and applications for Android devices. Design beautiful effects, animations, physical behaviors, and other different real-world features and techniques for your Android games and applications. Optimize your project and any other real-world projects for Android devices. Know more in practice about accessing Android functionality, rendering high-end graphics, expanding your project using asset bundles, and of course learn about deploying on the Android platform and much more.

Unity is essentially a development environment for multiple platforms (Android, iOS, Blackberry, Windows Phone, Windows, Playstation, Xbox, Mac, Linux, and Web). Unity is a very popular and effective technology for creating 2D and 3D games and applications. This technology provides many useful and effective tools to solve various issues. The Unity rendering engine provides great real-time rendering of high-quality graphics without too much cost and effort.

The Android platform and game industry is developing like never before. Most programmers using Unity, especially those new to the technology, would like to learn to recreate different functional parts from the most popular real-world games and applications that will cause a great concern in the community.

Whether you are new to Unity 5 or an expert, this book will provide you with the required skills you need to successfully design, implement, build, and enhance the quality of your Android game or application. By sequentially working through the steps in each chapter, you will quickly master the key features of the Unity 5 engine to implement real-world Android games and application features in practice.

This book is aimed for competent Unity developers who want to learn to develop, optimize, and publish games for Android devices. Know more about the Unity 5 engine to build awesome real-world Android games.

Starting from the beginning, this book will explain the configuration of the Unity 5 settings for the Android platform. Implement the innovative and user-friendly features using real-world techniques and tips and tricks in practice. Explore how to improve your Android games and enhance its performance. Open a wonderful world of rendering high-quality graphics with physically-based shaders and global illumination. Discover Android features inside the Unity 5 engine and more about transformation from native C# and JavaScript codes into Unity scripts.

This book is aimed at covering the fundamentals relating to typical real-world games and applications and to explore a basic overview of the concepts but the focus is on providing the practical skills required to develop Android games and applications.

Unity 5 for Android Essentials will teach you to use different tools provided by the Unity Technologies team in order to master your game's design and development processes. This book will be a practical guidebook that will help you to leverage the Unity 5 framework in order to build awesome games and apps for Android devices.

What this book covers

Chapter 1, Setting Up and Configuring an Android Platform, explains how to configure Unity 5 for Android devices. Also in this chapter, we will explore about APK expansion files in Unity 5. At the end of the chapter, you will build a very simple and basic game example for Android devices.

Chapter 2, Accessing Android Functionality, covers how to create plugins for the Android platform using Java and C languages in Unity 5. The reader will learn in practice how to write simple plugins for the Android platform. Also, the reader will learn to make anti-piracy checks, detect screen orientation, handle vibration support, determine device generation, and do more useful things.

Chapter 3, High-end Graphics for Android Devices, primarily explores how to enhance the quality in games and applications using physically-based shaders. This chapter will also describe global illumination in Unity 5. At the end of the chapter, you will optimize a shader code.

Chapter 4, Animation, Audio, Physics, and Particle Systems in Unity 5, will cover new Mecanim animation features in Unity 5. After that, you will learn awesome new audio features in Unity 5. At the end of this chapter, you will explore physics and particle systems in Unity 5.

Chapter 5, Asset Bundles in Unity 5 Pro, includes an overview of asset bundles in Unity 5. You will learn to download new code and data in real time for Android devices. At the end of this chapter, you will discover the safety technique of the asset bundles in practice.

Chapter 6, Optimization and Transformation Techniques, introduces the usage of occlusion culling and optimization techniques for the level of detail. You will learn to optimize Unity C# and Unity JS code. Finally, you will see how to transform Unity C# code to Unity JavaScript code and vice versa.

Chapter 7, Troubleshooting and Best Practices, covers the optimization of any game for Android devices and teaches you to find any bottlenecks. In this chapter, you will discover some troubleshooting techniques for the Android platform. At the end of this chapter, you will learn the best practices that are used by many professionals from all over the world for their scripts and shaders.

The online chapter, *Developing Glow Hockey from Scratch*, shows you how easy it is to develop the most popular game in the Android Market (Glow Hockey has about 100,000,000–500,000,000 installs `https://play.google.com/store/apps/details?id=com.natenai.glowhockey&hl=en`) in Unity 5 from scratch. Also, you will learn to optimize your project and any other real-world projects for Android devices. Many more useful details and features will be covered in this chapter. You can find this chapter at `https://www.packtpub.com/sites/default/files/downloads/9191OT_BonusChapter.pdf`.

What you need for this book

You need the following software for this book:

- Windows or Mac OS X
- Java Development Kit (JDK)
- Android SDK
- Unity3D

Who this book is for

This book is for readers that have a basic knowledge of the architecture of Unity, programming plus shading, and for those who are expert Unity developers. This book is very helpful for anybody who wants quick solutions and answers for many different questions and issues while developing a game or application in Unity.

Perhaps you already know a bit about the Unity 5 engine, but have never used it before; or perhaps, you know programming but are new to using Unity 5 to develop Android games and applications. In all cases, this book will quickly teach you to master high-quality Android games and applications. This book is for anyone who wants to explore a wide range of Unity features and finds ready-to-use techniques, solutions, tips and tricks that are used all over the world. It is best for you if you have basic experience with C# or JavaScript and feel comfortable enough with the Unity workflow.

Whether you've developed games before or not, and you're just looking to get started in Unity to make new fantastic real-world games and applications, this book will help you. This book is for you and everyone in your team, from beginners to expert developers!

Conventions

In this book, you will find a number of styles of text that distinguish between different kinds of information. Here are some examples of these styles and an explanation of their meaning.

Code words in text, database table names, folder names, filenames, file extensions, pathnames, dummy URLs, user input, and Twitter handles are shown as follows: "We can include other contexts through the use of the include directive."

A block of code is set as follows:

```
using UnityEngine;

public class YourClassName: MonoBehaviour {
  void OnCollisionExit (Collision collision) {
    Debug.Log ("OnCollisionExit :" + collision.gameObject.name);
  }
}
```

When we wish to draw your attention to a particular part of a code block, the relevant lines or items are set in bold:

```
using UnityEngine;

public class YourClassName: MonoBehaviour {
  void OnCollisionExit (Collision collision) {
    Debug.Log ("OnCollisionExit :" + collision.gameObject.name);
  }
}
```

Any command-line input or output is written as follows:

```
# cp /usr/src/asterisk-addons/configs/cdr_mysql.conf.sample
   /etc/asterisk/cdr_mysql.conf
```

New terms and **important words** are shown in bold. Words that you see on the screen, in menus or dialog boxes for example, appear in the text like this: "Clicking on the **Next** button moves you to the next screen."

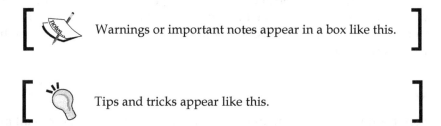

> Warnings or important notes appear in a box like this.

> Tips and tricks appear like this.

Reader feedback

Feedback from our readers is always welcome. Let us know what you think about this book—what you liked or may have disliked. Reader feedback is important for us to develop titles that you really get the most out of.

To send us general feedback, simply send an e-mail to feedback@packtpub.com, and mention the book title via the subject of your message.

If there is a topic that you have expertise in and you are interested in either writing or contributing to a book, see our author guide on www.packtpub.com/authors.

Customer support

Now that you are the proud owner of a Packt book, we have a number of things to help you to get the most from your purchase.

Downloading the example code

You can download the example code files for all Packt books you have purchased from your account at http://www.packtpub.com. If you purchased this book elsewhere, you can visit http://www.packtpub.com/support and register to have the files e-mailed directly to you.

Errata

Although we have taken every care to ensure the accuracy of our content, mistakes do happen. If you find a mistake in one of our books—maybe a mistake in the text or the code—we would be grateful if you would report this to us. By doing so, you can save other readers from frustration and help us improve subsequent versions of this book. If you find any errata, please report them by visiting http://www.packtpub.com/submit-errata, selecting your book, clicking on the **errata submission form** link, and entering the details of your errata. Once your errata are verified, your submission will be accepted and the errata will be uploaded on our website, or added to any list of existing errata, under the Errata section of that title. Any existing errata can be viewed by selecting your title from http://www.packtpub.com/support.

Piracy

Piracy of copyright material on the Internet is an ongoing problem across all media. At Packt, we take the protection of our copyright and licenses very seriously. If you come across any illegal copies of our works, in any form, on the Internet, please provide us with the location address or website name immediately so that we can pursue a remedy.

Please contact us at copyright@packtpub.com with a link to the suspected pirated material.

We appreciate your help in protecting our authors, and our ability to bring you valuable content.

Questions

You can contact us at questions@packtpub.com if you are having a problem with any aspect of the book, and we will do our best to address it.

1
Setting Up and Configuring an Android Platform

This chapter will talk about installing the Android SDK on Windows and Mac OS X platforms. Further, the reader will find how to configure Unity 5 for Android devices. Also, in this chapter, we will explore the APK expansion files in Unity 5. In this chapter, the reader will build the *Glow Hockey* project (we will create this game in Unity 5 from scratch in the last chapter in this book) on an Android device. At the end of this chapter, the reader will explore the side-by-side comparisons of the Unity Pro and Unity Basic-specific features and rules.

The topics that will be covered in this chapter are as follows:

- Configuring Unity 5 for Android devices
- APK expansion files in Unity 5
- Building for Android devices
- Unity license comparison overview

Configuring Unity 5 for Android devices

Once you have installed the Android SDK and set up Unity, you must configure the correct settings for each of your Android project. We will begin our review by considering the **Resolution And Presentation** options as shown in the following screenshot. In order to access the Android platform settings in Unity, it is necessary to navigate to the **Edit | Project Settings | Player** menu and then click on the button with the Android icon. Also, the other way of getting to the Android platform settings is by navigating to **File | Build Settings**. After opening the window, you will need to click on the **Player Settings** button at the bottom.

The **Default Orientation** option is shared between multiple mobile platforms. This setting is necessary in order to indicate the orientation of the screen which is designed for your game or your application. The default is **Automatic Rotation** for all settings. If, for example, your project is designed only for the portrait orientation of the screen, then you need to select either the **Portrait** or **Portrait Upside Down** value:

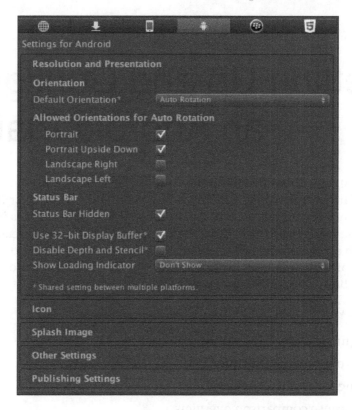

We can pick one of the following properties: **Portrait**, **Portrait Upside Down**, **Landscape Right**, **Landscape Left**, and **Auto Rotation**. They are pretty plain and speak for themselves. You just need to play a little with them to see their true purpose. The **Status Bar Hidden** checkbox needs no explanation because its meaning is obvious.

A subsequent option that we will survey is known as **Use 32-bit Display Buffer**. You can decide if the **display buffer** should handle 16-bit color values (if not 32-bit enabled) or if it should handle 32-bit color values. Remember that you need to activate this setting only if you have some artifacts, because it greatly affects the performance. The **Show Loading Indicator** field ensures the ensuing behaviors for us: **Don't Show**, **Large**, **Inversed Large**, **Small**, and **Inversed Small**.

As shown in the preceding figure, there are a lot of settings; however, most of them can be left with default values except for a few that must be adjusted before making a build. You cannot build an APK file for your Android device without configuring the **Bundle Identifier** option, which is shared between mobile platforms. The **Bundle Identifier** string must match the provisioning profile of the game you are building. The fundamental structure of the identifier is com.Company.ProductName. The bundle version is responsible for holding the number that describes the release index. Moreover, Unity permits us to specify the minimum API level that will be supported by your Android project. Also, you can set the name and icon for your application. Other settings are obvious and do not require additional explanation. More detailed information can be found in the official documentation of Unity.

APK expansion files in Unity 5

Google Play requires that the size of your games and applications does not exceed 50 MB. For most applications and games, this size is enough. Alternatively, you may want to have fantastic graphics for your projects and other huge media files that can take up a lot of space. Google Play makes a developer's life simpler and easier by expanding an APK file by large expansion files. Extension files are kept in a shared folder storage of the device where your game has enough access rights.

Overview

Each file cannot be larger than 2 GB, and you can choose any format for it. Of course, the best way is to use only compressed files in order to conserve bandwidth while the downloading process is active. You can add only one or two expansion files to your APK. Behind each expansion file lies its meaning:

- The first expansion file is known as the **main** and should be used for extra resources needed in your game. That's why this expansion file is primary.
- The second expansion file is known as **patch**, and is required to update the main file. That's why it is optional.

You should know that the Developer Console does not allow you to update your existing APK file by uploading only a new expansion file.

Formats

For your expansion files, you can use any desired format, such as MP3, MP4, AVI, RAR, ZIP, DOC, and PDF. The JOBB tool can help you encapsulate and encrypt your resources and patches for them.

The updating process

In most cases, Google Play will do all jobs automatically for you. So, very often, you don't need to do anything while users will download or upload your expansion files on their devices. However, sometimes your game has to download these files itself by receiving a URL from Google Play's application licensing service.

The basic steps to download expansion files for your game are listed as follows:

1. On the game start event, you should look for the expansion files in the `Android/obb/<package-name>/` directory.
2. In the first step, if you will find that your expansion files are already in that directory, then you can continue playing your game further.

3. In the event that the expansion files are not in that directory, you should perform the next two steps.

4. You have to receive URLs, names, and sizes for your game's expansion files. Before downloading anything, you should know where and what to download.

5. After having all the required information to download the expansion files, you can get your files and put them into the `Android/obb/<package-name>/` directory with the same name as Google Play told you.

> The following notes that are listed are taken from the official Android Developers Documentation page at `http://developer.android.com/google/play/expansion-files.html`:
>
> * The URL that Google Play provides for your expansion files is unique for every download, and each one expires shortly after it is given to your application.
> * Whether your application is free or not, Google Play returns the expansion file URLs only if the user acquired your application from Google Play.
> * A variety of errors may occur during the request and download that you must gracefully handle.
> * Network connectivity can change during the download, so you should handle such changes. If interrupted, resume the download when possible.
> * While the download occurs in the background, you should provide a notification that indicates the download progress, notifies the user when it's done, and takes the user back to your application when selected.

Setting up expansion files in Unity 5

Navigate to the **Player Settings | Publishing Settings** menu, and at the bottom you will see an option called **Split Application Binary**. When this option is enabled, your project will be divided into the `.apk` file for code, and for all other assets and data, it will be the `.obb` file.

Let's look at a list of key aspects related to the loading of the expansion files (`.obb`):

* The expansion files do not need to be uploaded to the server Google Play.

* If you have decided to publish `.apk` and `.obb` files on Google Play, then you need to include the code to download the expansion files.

- In the Unity Asset Store, you can find a great plugin for downloading and extracting your expansion files in the right location. The URL of this plugin is `http://u3d.as/content/unity-technologies/google-play-obb-downloader/2Qq`.

- Before testing the .obb files, you need to be logged in your Google account.

Building for Android devices

After creating a new project in Unity, it is a good idea to adjust the global quality settings as shown in the following figure. Most of them affect your game performance. Let's look deeper at **QualitySettings**, as it is a must before building applications for Android devices:

Unity allows you to create a template for your quality settings; you can also select one of the templates available in Unity by default. These settings greatly affect the performance of your application and the quality of your graphics. It is especially important for mobile platforms, where resources are very limited. You will need to play around with the settings on your target platforms to find the best template settings for your quality and performance. To access these settings, you need to navigate to **Edit | Project Settings | Quality**. You can select different templates separately for each platform supported by Unity. This setup window is divided into two main parts. The upper part, as shown in the preceding figure, is for managing templates, while the lower part, as shown in the following figure, is responsible for the settings themselves.

Each pattern (even Unity built-in templates) can be named as you wish. For all platforms supported by Unity, you can choose several accessible settings templates and also a default one. The default settings template is highlighted in green. Your settings should be as simple as possible, especially for mobile platforms. Unity allows you to create new settings templates and delete them by clicking on the icon with the basket.

The first part we intend to cover is **Rendering** as shown in the preceding screenshot. It contains the **Pixel Light Count** option that affects the upper limit pixel light only within **Forward Rendering** mode. The **Forward Rendering** path renders each object in one or more passes, depending on lights that affect the object. Lights themselves are also treated differently by **Forward Rendering**, depending on their settings and intensity.

The **Rendering** part contains the **Texture Quality** property with four existing options: **Full Res**, **Half Res**, **Quarter Res**, and **Eighth Res**. This lets you choose whether to display textures at maximum resolution or at a fraction of this (lower resolution has less processing overhead). Always remember, in any project you need to find a golden balance between the next two characteristics: quality and performance. The following property is named **Anisotropic Textures**, and it allows you to select just three values: **Disabled**, **Per Texture**, and **Forced On**. This describes if and how anisotropic textures will be used. On Wikipedia (http://en.wikipedia.org/wiki/Anisotropic_filtering), we can read next about anisotropic filtering: "In 3D computer graphics, **anisotropic filtering** (abbreviated **AF**) is a method of enhancing the image quality of textures on surfaces of computer graphics that are at oblique viewing angles with respect to the camera where the projection of the texture (not the polygon or other primitive on which it is rendered) appears to be non-orthogonal (thus the origin of the word: "an" for not, "iso" for same, and "tropic" from tropism, relating to direction; anisotropic filtering does not filter the same in every direction)".

The next property you learn about is **anti-aliasing**. Anti-aliasing can be turned off by selecting the **Disabled** option or it can be turned on by selecting **2x**, **4x**, and **8x Multi Sampling** options. The next setting is just to toggle soft blending for particles, and its name is **Soft Particles**. This was the last option for the **Rendering** part.

The next part is **Shadows**, and the name absolutely describes itself and its purpose. We can choose one value from three open values: **Hard and Soft Shadows**, **Hard Shadows Only**, and **Disable Shadows**. A big processing overhead can result if you choose the highest resolution for the **Shadow Resolution** option. Possible settings are the following: **Low**, **Medium**, **High**, and **Very High**. There are two separate routines for anticipating shadows from a directional light within the Shadow Projection option. If we choose **Close Fit**, then it renders higher resolution shadows, which can in some cases wobble marginally if the camera moves. The next option, and the last one for **Shadow Projection**, is the **Stable Fit** value, which is the opposite of **Close Fit**. This means that **Stable Fit** renders a lower resolution but without any artifacts when the camera moves. Next is the **Shadow Cascades** setting, which has an effect on processing overhead. A higher cycle of cascades can cope with more processing overhead. Do not forget about the golden balance. Obtainable options for cascades are next: **No Cascades**, **Two Cascades**, and **Four Cascades**.

http://docs.unity3d.com/460/Documentation/Manual/
DirectionalShadowDetails.html

On mobile platforms, real-time shadows for directional lights always use one shadow cascade and are **hard shadows**.

Directional lights are mostly used as a key light—sunlight or moonlight—in an outdoor game. Viewing distances can be huge, especially in first and third person games, and shadows often require some tuning to get the best quality versus performance balance for your situation.

The value of **Shadow Distance** is responsible for how far we can see shadows. Shadows beyond this length are visible, others are not.

In the following text, we are going to explore what is known as **Other** part. It contains five options for tuning any project. Let's start with the first option—**Blend Weights**. We can choose just three values which are very important for performance within this setting. The lower the value, the higher the performance. This setting tells the number of bones that can affect a given vertex during an animation. We can select **1 Bone**, **2 Bones**, or **4 Bones**; not more and not less. The next feature highly affects the performance and is not the first priority question about quality because of its hardly visible artifact. The name of this setting is **VSync Count** and the name of this artifact is known as **tearing**. If we want to avoid such artifacts then we need to synchronize rendering with the refresh rate of the display device, but do not forget about your performance. Synchronization can reduce a lot of your performance, so you should be ready for such kind of conditions. There are just three existing options for the **VSync Count** parameter: the first option is for synchronization with every **vertical blank (VBlank)**, the second value is for synchronization with every second vertical blank, and the third option allows us to disable all synchronizations, thereby speeding up your application. The next setting in our research is **LOD Bias**. This value does its work only when Unity needs to decide which LOD level to choose. For example, when there is a choice between two LOD levels, **LOD Bias** comes to help by selecting only one value. This is set in a range from zero to one as a fraction. The closer to zero, the less detailed a level will be chosen and vice versa. There are two remaining options that we will now consider. The first is **Maximum LOD Level**, and its purpose is to remember the number for the highest LOD level that you can use in your project. The second is **Particle RayCast Budget**, which needs particle system collisions with **Low** and **Medium** qualities, and this number describes the highest value of ray cast for physics approximation.

As for the **Maximum LOD level** parameter, all models in which the value will be less than this number will not be included in the build; Unity will ignore them, which can significantly reduce the amount of spending memory for your application or your game. The initial default value for this parameter is zero, which means that every model will be included in your build regardless of its level of detail. For each platform, depending on its configuration, Unity will use the smallest possible LOD level.

If the Android SDK installation and Unity setup are successful, you can safely create a build of your project. To do this, you will need to navigate to **File | Build Settings**, and in the opened window, you can create a build for various supported platforms as shown in the following figure. If you have properly installed the Android SDK and configured Android and have quality and player settings in Unity, you can safely click on the **Build** button or on the **Build And Run** button (if your Android device is properly configured and connected via USB) in the lower-right corner of the window.

Now, it's time to build Glow Hockey (we will create this game in Unity 5 from scratch in the last chapter in this book). First of all, you should create a new project in the Unity Editor. You can name it as you wish. To create a new project, you should click on the **Create Project** button in the bottom-right corner of the window.

The Unity Editor will be shown after creating the new project.

Glow Hockey is a very good example for the deployment of the project on the various platforms supported by Unity. In this game, there are many different effects, animations, sound effects, physics, and many other aspects from Unity. After deploying this project on an Android platform, you can test the various functions that are supported by Unity.

After opening the main **Glow Hockey** scene, you can make any changes or perform experiments as you wish. However, in this chapter, our goal is just to build this game on an Android device. We will not make any changes in the project in this chapter.

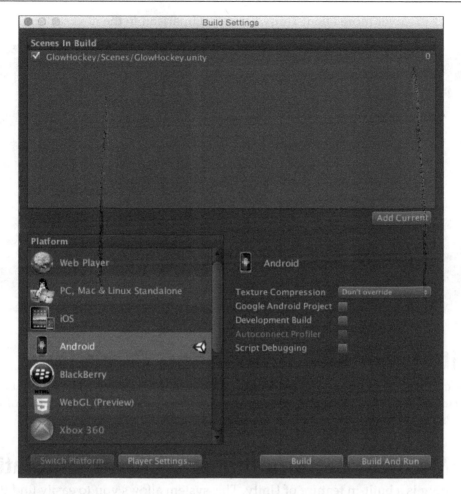

After opening the **Build Settings** window, you should select the **Android** platform, following which you can create an .apk file by clicking on the **Build** button in the bottom-right corner in order to share it with your friends, for example. Also, you can press the **Build And Run** button to export an .apk file as in the first case and to deploy this project right onto a connected device via the USB cable at the same time.

Now, let's discover more about Unity license comparison in the next section.

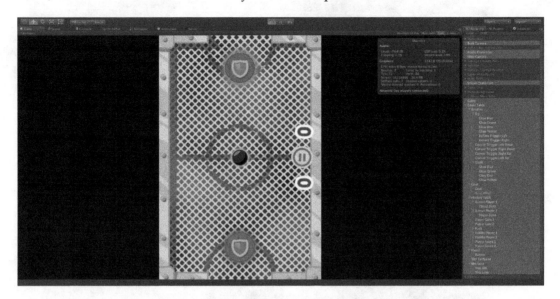

Unity License Comparison Overview

This section is based on the Unity License Comparison Overview. The link http://unity3d.com/unity/licenses will show you the side-by-side comparisons of the Unity Pro and Unity Basic's specific features and rules.

NavMeshes, pathfinding, and crowd simulation

Pathfinding is a built-in feature of Unity. This system allows you to easily find the right way from the beginning to the end, avoiding all the obstacles encountered on that way. All you have to do before you use this functionality is bake your navigation data in the Unity Editor. There, you must specify which floor objects or grounds can be walked on, and which objects are obstacles—all other problems will be solved by Unity without any effort after calling the pathfinding function with the start and end points as parameters. With a strong desire, you can create your own pathfinding system and crowd simulation. The pathfinding system in Unity is available for Basic and Pro licenses.

Level of Detail (LOD) support

Level Of Detail (LOD) allows you to optimize your productivity very well due to the fact that there are several different levels of quality for your mesh, each of which will be displayed in the camera view, depending on the distance to the camera. That is, when the camera is too far away, it is not optimal to display the most complex and detailed mesh, since all the power of all these details on the mesh will not be visible at all, which wastes precious resources. This is not very good for your performance as a whole. Detail mesh should be displayed only when the camera is close enough for you to see all these details on your mesh. LOD is supported only by the Unity Pro license. If you have only the Unity Basic license, then you can create your own LOD system very easily. The same idea for optimization can easily be implemented by yourself. The main thing is to change meshes in real time (more or less detailed) for your rendering, depending on their distance from the camera, which in turn will reduce unnecessary costs for the hardware.

Audio filters

Audio filters allow you to programmatically create different effects with the sound in real time. Imagine a situation where, in the game, you have to play a sound while your character is walking on sand. However, if the player suddenly gets into a tunnel, then the sound should be different from walking on sand. To solve this problem, you can choose one of the possible scenarios. The first solution to this problem lies in the fact that you can create or already use a ready-made sound for walking in tunnels and so on. For each situation, you will have a variety of ready-made sounds. Alternatively, this approach requires a lot of memory if you have too many different situations in the game where the game should play different sounds. This approach is not flexible relative to the second solution in the case of audio filters provided by Unity software in real time. Audio filters are supported only by the Unity Pro license.

Video playback and streaming

Today, many applications and games need to play different videos. A video may take a lot of memory, and this especially is an acute problem for mobile devices. In order to reduce the additional cost of memory for your video content, Unity allows you to broadcast via internet video streaming. This feature is only available on the Unity Pro license.

Full-fledged streaming with asset bundles

Asset bundles are supported only by the Unity Pro license. This feature greatly helps in optimizing the way of creating a quality game or application. This functionality allows developers to stream the content via the Internet, for example, adding new characters, new buildings, new weapons, new textures, and much more into the game.

Hundred thousand dollar turnover

This item is not a functionality in Unity, but it should rather be seen as a condition or requirement of Unity. This condition states that if in the last fiscal year, you (personally) or your organization, have earned more than $100,000 (inclusive), then you are required to use a Unity Pro license; that is, you have no right to use the Unity Basic license at all. This can be considered quite a logical and reasonable condition by Unity. After all, if you or your organization has earned more than $100,000 (or exactly $100,000), then you or your organization can buy a Unity Pro license without any problem.

Mecanim – IK Rigs

The new animation system in Unity, which is known as Mecanim, allows you to use a variety of different and useful functionalities, but a special opportunity and a key feature is the **Inverse Kinematics** (**IK**) Rigs. IK Rigs is supported in Mecanim for only humanoid characters with a correctly configured Avatar. The meaning of this feature is that you can call a function and pass a final point where the leg should be placed for hitting the ball (as one possible example, if you are creating a football game); after that, IK Rigs system will make all the rest automatically for you. For example, your character must take a cup on the table, but before that he will need to get up from the chair and go to the table on which the desired cup is, and only after that will your character be able to take a cup in his hand. These movements will play animations. All this hard work will completely rely on the IK Rigs system, you only need to specify the endpoint. IK Rigs is supported only by the Unity Pro license.

Mecanim – sync layers and additional curves

Mecanim also allows you to use different animation states simultaneously; for example, a character with full health will walk normally, but every time the health is decreased by about 20 percent, the character will begin to become worse and go slower, and then starts to limp at poor health. When the character's health becomes very poor, he begins to creep along the ground. This approach uses the sync layers option for grouping different animation states. This greatly simplifies the creation of a variety of conditions by the reusability of sync layers for different situations.

It is possible to modify sync layers dynamically in real time to reuse your state machine many times with different animations, but with the same conditions. Thus, developers do not need to create so many different state machines for all animations, but only a few, and reuse them while playing different animations. This feature is supported only by the Unity Pro license.

Additional curves allow you to add new curves to your animations in order to control different animation parameters. It's easy and very convenient to manage your animation curves in Unity Editor. This feature is supported only by the Unity Pro license.

Custom splash screen

This feature has the following meaning: while using the Unity Basic license, every time your application boots, your user sees the Unity logo image. If you want to replace that logo with your image, then you will need to purchase a Unity Pro license.

Build size stripping

This is a very important Unity feature, especially for mobile devices. With this feature, Unity allows you to remove all the excess out of your build. Unity helps you greatly in it, because it includes only those assets that are used in your game in your final build. Also, this feature allows you to include only those parts of the Unity engine that are utilized in your game in the final build. This feature is supported only by the Unity Pro license.

Lightmapping with global illumination and area lights

All Unity licenses support lightmapping. Unity allows you to bake lights and shadows for static objects. You can add more realism to your game by adding the global illumination and area lights provided by this feature, supported only by the Unity Pro license.

HDR and tone mapping

High Dynamic Range (HDR) and tone mapping functionality are very useful for improving the quality of the images in your game, but it requires a significant investment of resources. You must be very careful to use such an expensive operation, as well as many other expensive features in Unity. This feature allows you to use more colors than usual, which allows you to create, for example, morning light in the room. This feature is supported only by the Unity Pro license.

Occlusion culling

This feature is very useful for optimization. Unity excludes all unnecessary objects for rendering, such as those that are behind the wall or far from the camera. Otherwise, objects that are hiding will waste processor time and memory. You can easily create a system with the same idea for your specific tasks. This feature is supported only by the Unity Pro license.

Light probes

This functionality is used to supplement the lightmapping optimization method or the so-called light baking, which is used only for static objects, while dynamic objects look much worse. Light probes solve this problem for dynamic objects, but they must be used very carefully and gently so as not to harm the performance of your application or your game. This feature is supported only by the Unity Pro license.

Static batching

This functionality may optimize the rendering process in your game scene by reducing a large number of draw calls for static objects. This feature allows us to reduce many unnecessary draw calls. It works only for static objects, and is supported only by the Unity Pro license.

Render-to-texture effects

This Unity functionality is very interesting and often useful. This feature is useful when you want to directly render your camera not to the screen, but to your image. After that, you can do what you want with that image; for example, you can create a TV box in your game. Also, you can perform postprocessing effects with that image, and much more. However, this feature is very expensive, so use it carefully. This feature is supported only by the Unity Pro license.

Fullscreen postprocessing effects

This feature can also create very interesting effects. Alternatively, this functionality should be used very carefully, especially for mobile platforms as it can take a lot of resources for execution. While optimizing, you should not forget about its high price. For example, you can create effects such as a motion blur for a Formula 1 game, where cars go at a very high speed. Also, you can create bloom effects with this functionality, which makes objects glow like neon. This feature is supported only by the Unity Pro license.

NavMesh – dynamic obstacles and priority

While searching for the right way with a pathfinding system, there may be dynamic obstacles that your character should avoid. You can programmatically set objects as obstacles in your code for a certain time. The ability to manage priorities affects searching for the right path. This feature is supported only by the Unity Pro license.

.NET socket support

The ability to use the .NET sockets allows you to create a variety of network games, as well as connect directly to a device without a server. This feature is supported by both Unity Basic and Unity Pro licenses.

Profiler and GPU profiling

This is very useful for profiling your projects. Optimization should begin with finding bottlenecks in your application or in your game. To be more effective, while searching bottlenecks in your project, you should have good tools. You can create such tools yourself, or you can use ready-made solutions. One of the solutions provided by Unity is the profiler tool. This feature is supported only by the Unity Pro license. Those who have only a Unity Basic license have to create their own profiler tool. That's why at the end of the book, we will develop a very simple code profiler tool.

Real-time directional shadows

Lighting and shadows are key aspects in most games. Many developers from around the world create their games trying to achieve the most realistic lighting. No shadows in a world with realistic lighting is much worse than having shadows. Alternatively, to create such a realistic world requires a lot of resources, such as time and memory. This is especially important for mobile devices. You will need to find a balance between quality and performance. This feature is supported by both the Unity Basic and Unity Pro licenses.

Script access to asset pipeline

This feature is also very useful. With this functionality, you can automate the processing of large amounts of assets or builds. For example, imagine that you need to put a watermark on each of your textures. If it is only a few textures, then it can be done manually, but if there are too many textures, hundreds for example, then the automation of the processing of each picture will be very useful. For more information, you can look in the official documentation of Unity. Unity provides a variety of functions for the convenient handling of your assets and builds. This feature is supported only by the Unity Pro license.

Summary

In this chapter, we looked at how to install the Android SDK on Windows and Mac OS X. We also covered the various Unity settings before making the first build. After that, we looked at the APK expansion files in Unity for Android devices. Then, we talked about the build settings for Android. We created a very simple and small game build for Android platforms. At the end of the chapter, we considered step-by-step key points and the difference between the Unity Pro and Unity Basic licenses.

The next chapter has many more interesting details about the Android platform. You will learn how to create plugins within Unity for the Android platform. You will also find out how to make anti-piracy checks, detect screen orientation, handle vibration support, determine device generation, and many more useful things. Let's move on!

2
Accessing Android Functionality

In this chapter, you will discover how to create plugins for Android platform by Java and C languages in Unity 5. You will learn in practice how to write simple plugins for the Android platform. Also, the reader will learn how to carry out an anti-piracy check, detect screen orientation, handle vibration support, determine device generation, and other more useful features. The topics that will be covered in this chapter are as follows:

- Creating Java and native C plugins for an Android platform
- Android scripting API in Unity 5
- Accessing Android sensors and features within Unity 5

Creating Java and native C plugins for an Android platform

Before creating a Java or native C plugin for an Android platform inside Unity, you should install the Android NDK. Also, if you don't know how to build a shared library, then you should find out more about this process. A lot of information about Android NDK can be found on the web, for example, the Android official documentation can be found at `https://developer.android.com/tools/sdk/ndk/index.html` or in many different books by Packt Publishing; for example, see visit `https://www.packtpub.com/application-development/android-ndk-beginner's-guide`. Information about Android NDK is beyond the scope of this book.

Some parts of your game or app can be implemented using native-code languages, such as **C** or **C++**. The **Android NDK** is a toolset with multiple features and possibilities. You will not need to use the Android NDK in every project, but in some games or applications, it will be very helpful to reuse some custom or third-party code libraries in native languages, such as **C** or **C++**. There are much more possible use cases.

> Before using Android NDK, you should keep in mind that this approach is not always necessary and almost always increases the level of complexity of the code.

Creating plugins in C

Let's explore a very simple and basic plugin example that is written in C programming language as shown in the following code:

```
extern "C" {
  float Unity5AndroidPluginNativeC() {
    return 5.5f;
  }
}
```

After building our simple plugin example as a shared library, you should put it into the `Assets/Plugins/Android` directory. Now, let's discover how to use our native C plugin in Unity C# script as shown here:

```
[DLLImport ("NameOfYourPlugin")]
private static extern float Unity5AndroidPluginNativeC();
```

> Note that you cannot specify library extensions, such as `.lib` or `.so` in the name of your plugin. Also, you should wrap the entire native C code by Unity C# code, in order to check which platform your application is running on and if you can use this native C plugin.
>
> You have an opportunity to use precompiled Android libraries in Unity.

Let's look at how we can empower Unity 5, using the Java programming language for our Android games and applications. In order to use this advanced functionality in Unity, you must export Java code into a JAR file. Advanced functionality is not needed for each project, but this knowledge will still be useful for you often enough.

The Unity library for Android plugins is located on **Windows** at C:\
Program Files\Unity\Editor\Data\PlaybackEngines\
androidplayer\development\bin\classes.jar and C:\Program
Files\Unity\Editor\Data\PlaybackEngines\androidplayer\
release\bin\classes.jar.

The Unity library for Android plugins is located on **Mac OS X** at Unity/
Contents/PlaybackEngines/AndroidPlayer/development/
bin/classes.jar and Unity/Contents/PlaybackEngines/
AndroidPlayer/release/bin/classes.jar.

Creating plugins in Java (Eclipse IDE)

Next, let's take a look at how we can create our custom plugin by using Java
programming language in Eclipse IDE. You can also choose any other IDE which
will be comfortable for you. First, you need to create a new project as shown in the
following figure:

Once you choose **Android Application Project**, you should click on the **Next >** button at the bottom of the window. After that, you will see the window as shown in the screenshot here:

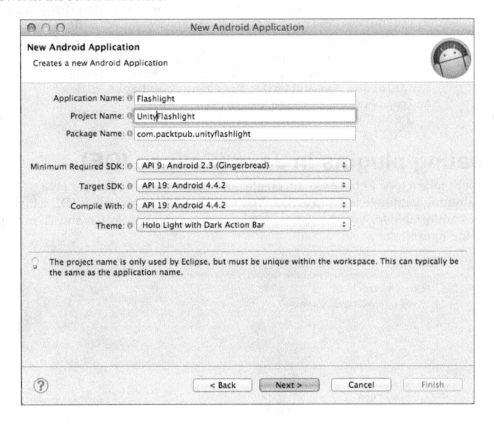

You should set the values for **Application Name**, **Project Name**, and **Package Name** as you wish or as in the preceding figure for our simple plugin example. Also, you can set other settings such as **Minimum Required SDK**, **Target SDK**, **Compile With**, and **Theme**. After that, click on the **Next >** button at the bottom of the window, and you will see the following window:

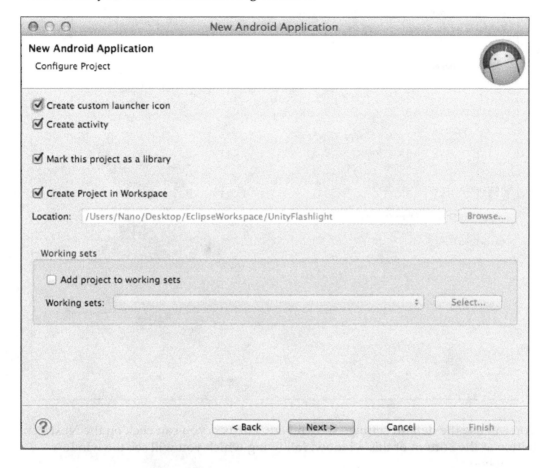

You can use the settings as shown in the preceding figure. After opening this window you should click on the **Next >** button at the bottom of this window, after which the following window will open:

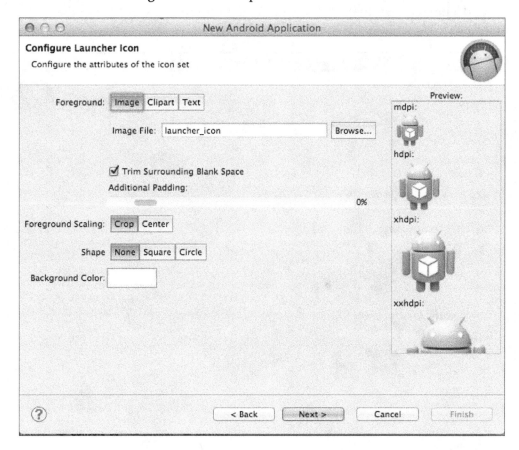

You can use the default settings shown here. Further, you can click on the **Next >** button at the bottom of this window, following which you will see the window shown here:

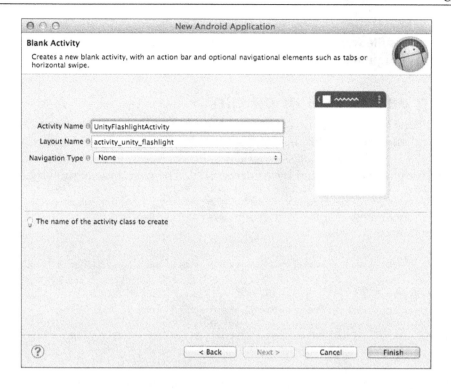

Here, you should set the **Activity name**, **Layout Name**, and **Navigation Type** fields as you wish or as seen in the earlier screenshot. Then, you can click on the **Finish** button at the bottom-right corner as shown in preceding figure.

Next, let's copy the Unity `classes.jar` library into the `libs` folder as shown here:

Now, open the `UnityFlashlightActivity.java` file shown in the preceding screenshot. This file was generated automatically by **Android Development Tools** (**ADT**, plugin for Eclipse IDE).

Writing Java code in plugin

This is the time to change your **UnityFlashlightActivity** code so that you can use this Android functionality in Unity scripts. For this, you should inherit the `UnityFlashlightActivity` class from **UnityPlayerActivity**, but not from the simple activity that is provided by Android SDK. The new code is shown here:

```
package com.packtpub.unityflashlight;
import com.unity3d.player.UnityPlayerActivity;
import android.os.Bundle;

public class UnityFlashlightActivity extends UnityPlayerActivity {

  @Override
  protected void onCreate(Bundle savedInstanceState) {
    super.onCreate(savedInstanceState);
  }
}
```

The next step is to create new the `Flashlight.java` class as shown here:

Firstly, we should declare our package name and the classes we need to import as shown in the code here:

```
package com.packtpub.unityflashlight;

import com.unity3d.player.UnityPlayerActivity;
import android.content.pm.PackageManager;
import android.hardware.Camera;
import android.hardware.Camera.Parameters;
```

Then, let's declare our `Flashlight` class and its variables as shown here:

```
public class Flashlight {
public UnityPlayerActivity unityPlayerActivity;
    public boolean isActiveFlashlight;

private Camera _cameraHardware;
```

The `cameraHardware` variable is the object of the `Camera` class that is provided by Android SDK, for using different features of the hardware camera. Now, it is time to write a constructor function for this class as shown here:

```
public Flashlight(UnityPlayerActivity upa) {
  // Unity Player Activity
unityPlayerActivity = upa;

  // Is Flashlight turned ON or OFF on the device
isActiveFlashlight = false;

// Receiving back hardware camera
_cameraHardware = Camera.open();
}
```

We will use the `unityPlayerActivity` variable for accessing the Android context and the `PackageManager` class that is provided by Android SDK. We will also use the `isActiveFlashlight` variable for turning the device's flashlight on and off. The last variable, `_cameraHardware`, will be used for accessing flashlight on a device. Now it is time to write a function that will check whether an Android device has the flashlight feature. The function code is shown in the following code:

```
public boolean HardwareHasFlashlight() {
        return (
unityPlayerActivity.
getPackageManager().
hasSystemFeature(PackageManager.FEATURE_CAMERA_FLASH)
        );
}
```

The next step is to describe the function that will turn on a flashlight on an Android device if it has this hardware feature:

```
public void ActivateFlashlight() {
if(HardwareHasFlashlight()) {
        isActiveFlashlight = true;

         _cameraHardware = Camera.open();

         Parameters params = _cameraHardware.getParameters();

         params.setFlashMode(Parameters.FLASH_MODE_TORCH);

         _cameraHardware.setParameters(params);

      // Turn ON a flashlight
         _cameraHardware.startPreview();
      }
}
```

Turning on and off a hardware flashlight

Now let's write a function that will turn off a flashlight on an Android device, if it has this hardware feature:

```
public void DeactivateFlashlight() {
if(HardwareHasFlashlight()) {
       isActiveFlashlight = false;

    // Turn OFF a flashlight
       _cameraHardware.stopPreview();

       _cameraHardware.release();
   }
}
```

The simplest method in this class is shown in the following code. Next, we should close our `Flashlight` class using the following code:

```
public boolean IsActiveFlashlight() {
       return isActiveFlashlight;
   }
}
```

If you receive an error with the `unityPlayerActivity.getPackageManager()` function, then you will need to change your minimal Android SDK version in the `AndroidManifest.xml` file as shown in the line of code here:

```
<uses-sdk android:minSdkVersion="9"/>
```

Now, let's change our UnityFlashlightActivity a little, as shown in the following code:

```
package com.packtpub.unityflashlight;

import com.unity3d.player.UnityPlayerActivity;

import android.os.Bundle;

public class UnityFlashlightActivity extends UnityPlayerActivity {
  public Flashlight flashlight = new Flashlight(this);

  @Override
  protected void onCreate(Bundle savedInstanceState) {
    super.onCreate(savedInstanceState);
  }
}
```

On Unity side

Here, we will create a new scene for our Android plugin test in our project:

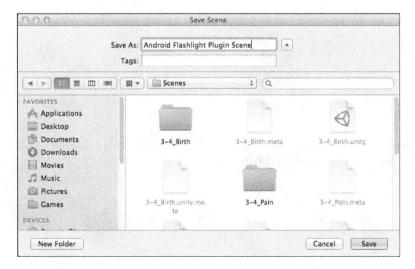

Name this new scene as you wish, following which you should create a new `Assets/`
`Plugins/Android` folder. You can put in the `AndroidManifest.xml`, JAR files, and
Android resources files in this folder.

Exporting and importing a JAR library from Eclipse into Unity

Now, let's go back to the Eclipse editor and right click on the mouse button on our
Flashlight project. Click on the **Export...** button as shown here:

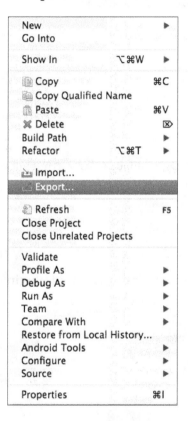

You will see the window as shown in the following screenshot. Select **JAR file** and
click on the **Next >** button at the bottom of the window.

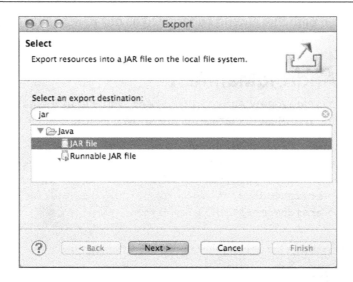

In the following window, select the options you wish or just set them up as shown here:

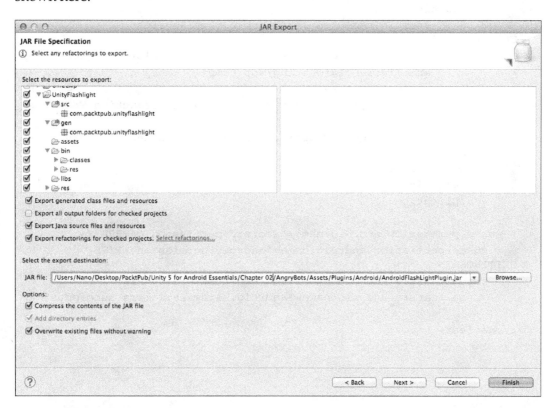

Choose the JAR file path that should be targeted as our new `Asset/Plugins/ Android` folder in Unity.

Importing AndroidManifest

The next step is to create a new `AndroidManifest.xml` file in the `Asset/Plugins/ Android` folder. The manifest declarations are shown as follows:

```xml
<?xml version="1.0" encoding="utf-8"?>
<manifest xmlns:android="http://schemas.android.com/apk/res/android"
    package="com.packtpub.unityflashlight"
    android:versionCode="1"
    android:versionName="1.0" >

    <uses-sdk android:minSdkVersion="9"/>

    <application
        android:icon="@drawable/app_icon"
        android:label="@string/app_name">
        <activity
            android:name="com.packtpub.unityflashlight.
UnityFlashlightActivity"
            android:configChanges = "keyboardHidden|orientation"
            android:label="@string/app_name" >
            <intent-filter>
                <action android:name="android.intent.action.MAIN" />

                <category android:name="android.intent.category.
LAUNCHER" />
            </intent-filter>
        </activity>
    </application>

  <uses-permission android:name="android.permission.CAMERA"/>
    <uses-permission android:name="android.permission.WRITE_
SETTINGS"/>
    <uses-feature android:name="android.hardware.camera" />
    <uses-feature android:name="android.hardware.camera.autofocus" />

</manifest>
```

Using Java plugins in Unity scripts

Here, we will create a very simple `FlashlightActivity.cs` script in Unity as shown in the following code:

```
using UnityEngine;

public static class FlashlightActivity
{
  #if UNITY_ANDROID && !UNITY_EDITOR
    public static AndroidJavaClass activityClass =
new AndroidJavaClass("com.packtpub.unityflashlight.
UnityFlashlightActivity");
    public static AndroidJavaClass unityActivityClass = new
AndroidJavaClass("com.unity3d.player.UnityPlayer");
    public static AndroidJavaObject activityObj = unityActivityClass.
GetStatic<AndroidJavaObject>("currentActivity");
  #else
    public static AndroidJavaClass activityClass;
    public static AndroidJavaClass unityActivityClass;
    public static AndroidJavaObject activityObj;
  #endif
}
```

Then, we will implement `Flashlight.cs` as shown here:

```
using UnityEngine;

public static class Flashlight
{
  #if UNITY_ANDROID && !UNITY_EDITOR
    private static AndroidJavaObject flashlight = FlashlightActivity.
activityObj.Get<AndroidJavaObject>("flashlight");
  #else
    private static AndroidJavaObject flashlight;
  #endif

  public static bool HardwareHasFlashlight()
  {
    if (Application.platform == RuntimePlatform.Android)
      return flashlight.Call<bool>("HardwareHasFlashlight");
    else
      return false;
  }
```

```
public static bool IsActiveFlashlight()
{
  if (Application.platform == RuntimePlatform.Android)
    return flashlight.Call<bool>("IsActiveFlashlight");
  else
    return false;
}

public static void ActivateFlashlight()
{
  if (Application.platform == RuntimePlatform.Android)
    flashlight.Call("ActivateFlashlight");
}

public static void DeactivateFlashlight()
{
  if (Application.platform == RuntimePlatform.Android)
    flashlight.Call("DeactivateFlashlight");
}
}
```

This class is very simple, and we will not explain it. This class just recalls the Android Java methods. In the end, let's create one more very simple class that should be attached on **MainCamera** in our new scene. The filename is FlashlightTest.cs, and its code is shown here:

```
using UnityEngine;

public class FlashlightTest : MonoBehaviour {
  void Start () {
    Flashlight.ActivateFlashlight();
  }

  void OnApplicationQuit() {
    Flashlight.DeactivateFlashlight();
  }
}
```

At the end, we can make a new build for Android devices with just one single scene as shown in the following screenshot:

We can access many different Android properties ready for use in Unity 5. You should use the UNITY_ANDROID defined constant by Unity for conditionally, compiling the Android-specific C# code. Further, in this chapter you will learn how to use different Android features and the properties inside Unity scripts. Most of the Android features in Unity are presented by **Handheld** and **Input** classes. Let's look at the Android features you will explore next:

- Anti-piracy check
- Vibration
- Activity indicator
- Screen orientation
- System information

Anti-piracy check

Let's explore **anti-piracy** checks first. Very often, if not always, you should protect your Android game or application from pirates that hack games and applications in order to redistribute them for free. With the help of Unity, we can make an anti-piracy check that shows if our game or application was changed after it was built.

You should check the `Application.genuine` Boolean property, which is provided by the Unity library. If this property is `false`, then you can notify the user that it is a hacked version, or you can cut some functionality, and you can do any other actions or combinations of them. The `Application.genuine` check is a very expensive operation, so you it is best not to check this property so often, just when it is required.

Vibration

In order to make an Android device vibrate, you should call the `Handheld.Vibrate()` method, which is provided by the Unity library.

Activity Indicator

You can use **Activity Indicator** for slow operations. There is an Android built-in Activity Indicator. Let's explore a very simple code example which is shown here:

```
using UnityEngine;
using System.Collections;

public class ShowActivityIndicator : MonoBehaviour {
  IEnumerator ActivityIndicatorExample()
  {
    #if UNITY_ANDROID
      Handheld.SetActivityIndicatorStyle(
        AndroidActivityIndicatorStyle.Small
      );
    #endif

    Handheld.StartActivityIndicator();
    yield return new WaitForSeconds(0);
    Application.LoadLevel(1);
  }

  void OnGUI()
  {
```

```
    if( GUI.Button(new Rect(50, 50, 300, 300), "Start") ) {
      StartCoroutine(ActivityIndicatorExample());
    }
  }
}
```

Screen orientation

Also, you can control the Android screen orientation inside Unity scripts. You can detect a screen rotation or you can force a screen rotation to a specific orientation. In order to get the current device orientation, you should access the `Screen.orientation` property. Also, you can set this property to any desired screen orientation to force rotation. You can find more about different screen orientation properties and constants in the Unity manual.

System information

If you need more information about your system, you can use static variables from the `SystemInfo` class, which is provided by the Unity library. More information about these variables can be found in the official Unity documentation located at `http://docs.unity3d.com/ScriptReference/SystemInfo.html`.

You can attach the following script in any scene on any object in order to obtain information about your Android device. Let's look more closely at what happens inside this code. Firstly, we need to declare our class with its properties as shown here:

```
using UnityEngine;

public class ShowSystemInfo : MonoBehaviour {
  public Vector2 scrollPosition;

  private Vector2 _v1, _v2;
```

After that let's create our `OnGUI` function, which will show us information about any device. The code is shown here:

```
  void OnGUI()
  {
    GUILayout.BeginVertical();

    scrollPosition = GUILayout.BeginScrollView(
      scrollPosition,
      GUILayout.Width(Screen.width),
```

```
        GUILayout.Height(Screen.height)
    );


    GUILayout.Label("SystemInfo.deviceModel <<<===>>> " + SystemInfo.
deviceModel);
    GUILayout.Label("SystemInfo.deviceName <<<===>>> " + SystemInfo.
deviceName);
    GUILayout.Label("SystemInfo.deviceType <<<===>>> " + SystemInfo.
deviceType.ToString());
    GUILayout.Label("SystemInfo.deviceUniqueIdentifier <<<===>>> " +
SystemInfo.deviceUniqueIdentifier);
    GUILayout.Label("SystemInfo.graphicsDeviceID <<<===>>> " +
SystemInfo.graphicsDeviceID.ToString());
    GUILayout.Label("SystemInfo.graphicsDeviceName <<<===>>> " +
SystemInfo.graphicsDeviceName);
    GUILayout.Label("SystemInfo.graphicsDeviceVendor <<<===>>> " +
SystemInfo.graphicsDeviceVendor);
    GUILayout.Label("SystemInfo.graphicsDeviceVendorID <<<===>>> " +
SystemInfo.graphicsDeviceVendorID.ToString());
    GUILayout.Label("SystemInfo.graphicsDeviceVersion <<<===>>> " +
SystemInfo.graphicsDeviceVersion);
    GUILayout.Label("SystemInfo.graphicsMemorySize <<<===>>> " +
SystemInfo.graphicsMemorySize.ToString());
    GUILayout.Label("SystemInfo.graphicsPixelFillrate <<<===>>> " +
SystemInfo.graphicsPixelFillrate.ToString());
    GUILayout.Label("SystemInfo.graphicsShaderLevel <<<===>>> " +
SystemInfo.graphicsShaderLevel.ToString());
    GUILayout.Label("SystemInfo.npotSupport <<<===>>> " + SystemInfo.
npotSupport.ToString());
    GUILayout.Label("SystemInfo.operatingSystem <<<===>>> " +
SystemInfo.operatingSystem);
    GUILayout.Label("SystemInfo.processorCount <<<===>>> " +
SystemInfo.processorCount.ToString());
    GUILayout.Label("SystemInfo.processorType <<<===>>> " +
SystemInfo.processorType);
    GUILayout.Label("SystemInfo.supportedRenderTargetCount <<<===>>> "
+ SystemInfo.supportedRenderTargetCount.ToString());
    GUILayout.Label("SystemInfo.supports3DTextures <<<===>>> " +
SystemInfo.supports3DTextures.ToString());
    GUILayout.Label("SystemInfo.supportsAccelerometer <<<===>>> " +
SystemInfo.supportsAccelerometer.ToString());
    GUILayout.Label("SystemInfo.supportsComputeShaders <<<===>>> " +
SystemInfo.supportsComputeShaders.ToString());
    GUILayout.Label("SystemInfo.supportsGyroscope <<<===>>> " +
SystemInfo.supportsGyroscope.ToString());
```

```
    GUILayout.Label("SystemInfo.supportsImageEffects <<<===>>> " +
SystemInfo.supportsImageEffects.ToString());
    GUILayout.Label("SystemInfo.supportsInstancing <<<===>>> " +
SystemInfo.supportsInstancing.ToString());
    GUILayout.Label("SystemInfo.supportsLocationService <<<===>>> " +
SystemInfo.supportsLocationService.ToString());
    GUILayout.Label("SystemInfo.supportsRenderTextures <<<===>>> " +
SystemInfo.supportsRenderTextures.ToString());
    GUILayout.Label("SystemInfo.supportsRenderToCubemap <<<===>>> " +
SystemInfo.supportsRenderToCubemap.ToString());
    GUILayout.Label("SystemInfo.supportsShadows <<<===>>> " +
SystemInfo.supportsShadows.ToString());
    GUILayout.Label("SystemInfo.supportsSparseTextures <<<===>>> " +
SystemInfo.supportsSparseTextures.ToString());
    GUILayout.Label("SystemInfo.supportsStencil <<<===>>> " +
SystemInfo.supportsStencil.ToString());
    GUILayout.Label("SystemInfo.supportsVibration <<<===>>> " +
SystemInfo.supportsVibration.ToString());
    GUILayout.Label("SystemInfo.systemMemorySize <<<===>>> " +
SystemInfo.systemMemorySize.ToString());

    GUILayout.EndScrollView();
    GUILayout.EndVertical();
  }
```

At the end, we should create the Update function in order to change the scroll position Y value, in order to be able to scroll the list of information:

```
void Update() {
    if (Input.touchCount > 0) {
        if (TouchPhase.Began == Input.GetTouch(0).phase) {
            _v1 = _v2 = Input.GetTouch(0).position;
        } else if (TouchPhase.Moved == Input.GetTouch(0).phase) {
            _v2 = _v1;

            _v1 = Input.GetTouch(0).position;

            scrollPosition.y += (_v1.y > _v2.y ? -1 : 1) * Vector2.
Distance(_v1, _v2);
        }
    } else {
        if (Input.GetMouseButtonDown(0)) {
            _v1 = _v2 = new Vector2(Input.mousePosition.x, Input.
mousePosition.y);
```

```
    } else if (Input.GetMouseButton(0)) {
      _v2 = _v1;

      _v1 = new Vector2(Input.mousePosition.x, Input.
mousePosition.y);

      scrollPosition.y += (_v1.y > _v2.y ? -1 : 1) * Vector2.
Distance(_v1, _v2);
    }
   }
  }
}
```

Accessing the Android sensors and features within Unity 5

Let's discover a little bit more about some Android sensors and features that are shown in the simple code examples later in text.

Acceleration

You can also access the Android acceleration inside Unity. Let's upgrade our previous example, so we can see one more aspect of our Android device. The new piece of code is shown here:

```
GUILayout.Label("\n\n A C C E L E R A T I O N");
    GUILayout.Label("Input.acceleration = (" + Input.acceleration.x +
", " + Input.acceleration.y + ", " + Input.acceleration.z + ")");
```

After running this test on your Android device, you will see how acceleration values change very quickly in real time.

Gyroscope

Unity provides **gyroscope** access on Android devices as shown in the new piece of code. Before using gyroscope, you just need to enable it as shown here:

```
    Input.gyro.enabled = true;

GUILayout.Label("\n\n G Y R O S C O P E");
    GUILayout.Label("Input.gyro.attitude <<<===>>> " + Input.gyro.
attitude.ToString());
```

```
        GUILayout.Label("Input.gyro.enabled <<<===>>> " + Input.gyro.
enabled.ToString());
        GUILayout.Label("Input.gyro.gravity <<<===>>> " + Input.gyro.
gravity.ToString());
        GUILayout.Label("Input.gyro.rotationRate <<<===>>> " + Input.gyro.
rotationRate.ToString());
        GUILayout.Label("Input.gyro.rotationRateUnbiased <<<===>>> " +
Input.gyro.rotationRateUnbiased.ToString());
        GUILayout.Label("Input.gyro.updateInterval <<<===>>> " + Input.
gyro.updateInterval.ToString());
        GUILayout.Label("Input.gyro.userAcceleration <<<===>>> " + Input.
gyro.userAcceleration.ToString());
```

Compass

Unity provides **compass** access on Android devices as shown in the new piece of
code. Before using the compass, you just need to enable it as follows:

```
    Input.compass.enabled = true;


    GUILayout.Label("\n\n C O M P A S S");
    GUILayout.Label("Input.compass.enabled <<<===>>> " + Input.
compass.enabled.ToString());
    GUILayout.Label("Input.compass.headingAccuracy <<<===>>> " +
Input.compass.headingAccuracy.ToString());
    GUILayout.Label("Input.compass.magneticHeading <<<===>>> " +
Input.compass.magneticHeading.ToString());
    GUILayout.Label("Input.compass.rawVector <<<===>>> " + Input.
compass.rawVector.ToString());
    GUILayout.Label("Input.compass.timestamp <<<===>>> " + Input.
compass.timestamp.ToString());
    GUILayout.Label("Input.compass.trueHeading <<<===>>> " + Input.
compass.trueHeading.ToString());
```

We can see all values received from our previous code examples in the following screenshot:

SystemInfo.deviceModel <<<===>>> samsung SM-G900H
SystemInfo.deviceName <<<===>>> <unknown>
SystemInfo.deviceType <<<===>>> Handheld
SystemInfo.deviceUniqueIdentifier <<<===>>> 0456f762d8193fbe4c71921de1e2b600
SystemInfo.graphicsDeviceID <<<===>>> 0
SystemInfo.graphicsDeviceName <<<===>>> Mali-T628
SystemInfo.graphicsDeviceVendor <<<===>>> ARM
SystemInfo.graphicsDeviceVendorID <<<===>>> 0
SystemInfo.graphicsDeviceVersion <<<===>>> OpenGL ES 3.0
SystemInfo.graphicsMemorySize <<<===>>> 252
SystemInfo.graphicsPixelFillrate <<<===>>> -1
SystemInfo.graphicsShaderLevel <<<===>>> 30
SystemInfo.npotSupport <<<===>>> Full
SystemInfo.operatingSystem <<<===>>> Android OS 4.4.2 / API-19 (KOT49H/G900HXXU1ANG3)
SystemInfo.processorCount <<<===>>> 8
SystemInfo.processorType <<<===>>> ARMv7 VFPv3 NEON
SystemInfo.supportedRenderTargetCount <<<===>>> 1
SystemInfo.supports3DTextures <<<===>>> False
SystemInfo.supportsAccelerometer <<<===>>> True
SystemInfo.supportsComputeShaders <<<===>>> False
SystemInfo.supportsGyroscope <<<===>>> True
SystemInfo.supportsImageEffects <<<===>>> True
SystemInfo.supportsInstancing <<<===>>> False
SystemInfo.supportsLocationService <<<===>>> True
SystemInfo.supportsRenderTextures <<<===>>> True
SystemInfo.supportsRenderToCubemap <<<===>>> True
SystemInfo.supportsShadows <<<===>>> True
SystemInfo.supportsSparseTextures <<<===>>> False
SystemInfo.supportsStencil <<<===>>> 1
SystemInfo.supportsVibration <<<===>>> True
SystemInfo.systemMemorySize <<<===>>> 1796

ACCELERATION

Input.acceleration = {-0.08880615, -0.8681641, -0.4262696}

GYROSCOPE

Input.gyro.attitude <<<===>>> (0.3, 0.4, 0.7, 0.5)
Input.gyro.enabled <<<===>>> True
Input.gyro.gravity <<<===>>> (-0.1, -0.9, -0.4)
Input.gyro.rotationRate <<<===>>> (-0.1, 0.0, 0.0)
Input.gyro.rotationRateUnbiased <<<===>>> (-0.1, 0.0, 0.0)
Input.gyro.updateInterval <<<===>>> 2E+10
Input.gyro.userAcceleration <<<===>>> (0.0, 0.0, 0.0)

COMPASS

Input.compass.enabled <<<===>>> True
Input.compass.headingAccuracy <<<===>>> 0
Input.compass.magneticHeading <<<===>>> 233.578
Input.compass.rawVector <<<===>>> (9.0, -30.0, -8.3)
Input.compass.timestamp <<<===>>> 1.40982335984297E+18
Input.compass.trueHeading <<<===>>> 233.578

Summary

In this chapter, we look at the details of writing plugins in Java and native C for Unity. We practiced developing simple plugins for Android platform. Also, we explored how to make anti-piracy checks, detect screen orientation, handle vibration support, get device name, get device model, and get much more useful information. We learned how to access the acceleration, gyroscope, and compass sensors with their features and properties in practice.

In the next chapter, you will learn how to develop high-end graphics. You will explore how to write simple Cg shaders in Unity. You will also learn more about awesome tips, tricks, and techniques that are used all over the world in game production. You will also acquire information about global illumination and how to optimize your shaders.

3

High-end Graphics for Android Devices

Primarily, this chapter will explore how to enhance the quality of games and applications using different techniques and physically-based shaders. In this chapter, firstly we will examine the different techniques of lighting that are often used in game development and production stages. Secondly, this chapter will describe global illumination in Unity 5. At the end of the chapter, the reader will optimize a shader code.

The topics that will be covered in the chapter are as follows:

- Physically-based shaders
- Global illumination
- Practicing in shader optimization

Physically-based shaders

Unity makes it very easy to use ready shaders or write your own shaders using the Cg language or the surface shaders framework. Surface shaders are written in Cg, but they do a large amount of work that you do not have to write every time when you create a new shader. The surface shader language uses a component-based approach, or in other words a more abstract approach that facilitates writing complex shaders using a sophisticated lighting model. While using the surface shaders framework, graphics programmers do not have to keep reprocessed texture coordinates and matrix transformations. In this chapter, we will describe the different techniques and methods of writing performance-friendly shaders in great detail with nice visual quality effects used in a variety of games and applications developed throughout the world.

First, let's start with the basic principles and concepts of a shaders. A shader is a pre-compiled program for one of the number of stages of the graphics pipeline used in three-dimensional graphics to determine the final parameters of the object or image. It may include a description of arbitrary complexity absorption and scattering of light, texture mapping, reflection and refraction, shading, surface displacement, and postprocessing effects.

Basic shader concepts

Programmable shaders are flexible and effective. Seemingly complex surfaces can be visualized with simple geometric forms. For example, the shaders can be used to draw a three-dimensional surface of the ceramic tiles on a completely flat surface.

In Unity, shaders are divided into three types: **vertex**, **geometry**, and **fragment (pixel)**.

The vertex shader

The **vertex shader** manipulates data mapped to the vertices of polyhedra. This data corresponds to the coordinates of the vertices in space, texture coordinates, a tangent vector, a bi-normal vector, and a normal vector. The vertex shader can be used for perspective vertices transformation, for generating texture coordinates, for lighting calculations, and so on.

The geometry shader

The **geometry shader**, in contrast to the vertex shader, is able to handle not only one vertex, but also a whole primitive with a set of vertexes (triangles, quads, and so on). It can be cut (two vertices) and triangular (three vertices), and the information on adjacent vertices (adjacency) can be processed for up to six vertices of the triangular primitive. Furthermore, the geometry shader can generate primitives "on the fly," without using a central processor. They were first used on the *Nvidia 8* series.

The pixel/fragment shader

The **pixel shader** works with fragments of the bitmap. A pixel shader is used in the last stage of the graphics pipeline to generate a fragment of the picture.

Shading languages

Shading languages usually contain special data types such as matrices, samplers, vectors, and a set of built-in variables and constants for easy integration with different 3D libraries. As computer graphics have many application areas, to meet the different needs of the market, developers created a large number of shader languages.

These shading languages are focused on delivering the highest quality image. Describing the properties of materials made at the highest abstract level to work do not need any special skills or knowledge of programming hardware. Such shaders are usually created by artists in order to ensure "the right kind" such as texture mapping, light sources, and other aspects of art and science at the same time.

Processing these shaders is usually a resource-intensive task. The aggregate computing power needed for this work can be very high, as it is used to create photo-realistic images. The main part of the similar computation is performed by a visualization of large computer clusters.

Cg

The Cg shading language developed by Nvidia in conjunction with Microsoft (essentially the same language from Microsoft is called **HLSL**, and is included in DirectX 9). Cg is used in Unity and stands as **C for Graphics**. The language is very similar to C and it uses similar data types (int, float, and a special 16-bit floating point type—half). Cg also supports functions and structures. The language has peculiar optimizations like packed arrays—type declarations like float a [4] and float4 a are different types. The second announcement is a **packed array**. The packed array operations are faster than conventional operations. Despite the fact that the language was developed by Nvidia, it works without problems with other graphics cards (for example, ATI cards). However, please note that all the shader programs have their own peculiarities, which can be obtained from specialized sources.

Unity shaders in Cg

In addition, you should know that Unity 5 comes with built-in shaders, which are very useful, especially for some basic stuff required in many different games. Now let's start our wonderful journey into the world of Unity shaders in Cg language. Typically, shaders use diffuse components or a lighting model. First, you must understand what should be optimized in your shaders well. Basically, you should try to avoid complicated calculations and labor-intensive functions. In this chapter, firstly, we will examine the different techniques of lighting models that are so often used in game development and production stages. Lighting is one of the fundamental aspects of the shader. Therefore, programmers often use their approximate calculations for lighting to speed up performance.

Earlier, computer graphics were used as a fixed-function lighting model, which was not a very flexible solution, since it gave graphics programmers only a single lighting model that could only be adjusted by setting a finite set of parameters and textures. Unlike before, where a single fixed-lighting model was used, today, developers use a very flexible programmable approach to create different lighting models with the help of the Cg shader language, especially from the wonderful surface shaders in Unity.

The diffuse component in the shader will often specify exactly how the rays of light reflect from the surface in all directions. You may find that it is very similar to the work of a mirror that reflects the sun's rays at different angles and in all directions. However, it is not so, and we'll show you this difference in as much detail as possible later on in the chapter.

The main difference is that a reflective surface like a mirror reflects the image of the surrounding environment, while the diffuse lighting model reflects sunlight back into the field of view.

In order to create a simple and basic diffuse lighting model, you will need to create a shader that will contain an emissive color, an ambient color, and of course the total accumulation of color from all light sources. Techniques and tricks that we will show you in the following code will help you create your own lighting models, as well as explore various industry tricks to help you understand the basic ideas to create more complex lighting models using only the textures that will give a huge increase to your productivity. In other words, the use of premade textures to create lighting models can greatly increase your productivity.

Let's start with the simplest example of our surface shader as shown in the following code. The code was generated by the Unity Editor:

```
// The first line of our shader code specifies the name of the
// shader in order to further select it from a list of all
// shaders.
Shader "PacktPub/SimpleDiffuseLighting"
{
  // Next is the properties block of parameters of the shader
  // known as Properties, which is followed by a block of the
  // shader code known as SubShader.
      Properties
      {
              _MainTex ("Base (RGB)", 2D) = "white" {}
      }
```

```
SubShader
{
        Tags {"RenderType" = "Opaque"}
        LOD 200

        CGPROGRAM
        #pragma surface surf Lambert

        sampler2D _MainTex;
        struct Input
        {
                float2 uv_MainTex;
        };

        void surf (Input IN, inout SurfaceOutput o)
        {
                half4 c = tex2D (_MainTex, IN.uv_MainTex);
                o.Albedo = c.rgb;
                o.Alpha = c.a;
        }
        ENDCG
    }

// The shader specified as FallBack will be executed
// instead of our shader.
    FallBack "Diffuse"
}
```

Let's consider the `Properties` block in more detail. The properties are some of the very important elements while writing shaders. The properties allow the artist to set their own textures or other settings to customize the visual effects. You can tune the properties of the selected shader through Unity materials.

Unity parses each shader code in order to find built-in structures. The `Properties` block is one of these built-in structures that Unity is looking for. Here is an example of the structure of the `Properties` block:

```
Properties
{
        _YourVariableName ("Inspector GUI Name", Color) = (1,1,1,1)
}
//      Variable Name    Inspector GUI Name    Type  Default Value
```

Each time you create a new property, you will need to name your variable. The variable name is used in the code of your shader, while the inspector GUI name will be shown in the Unity Editor. The type can be any one of the following:

- `Range (min, max)`: These are real numbers in the form of a slider from `min` to `max`
- `Color`: This opens the color picker in the Unity inspector for choosing the desired color value
- `2D`: This is used for adding textures
- `Rect`: This is a nonpower of two textures
- `Cube`: This is a cube map texture
- `Float`: These are real values without the slider
- `Vector`: This is a four-component vector with real numbers

At the end of the `Properties` structure, we specified the default value.

A custom diffuse lighting model

Before you write your own diffuse lighting model, we will consider our new properties:

```
Properties
{
        _FirstColor ("First Color", Color) = (1,1,1,1)
        _SecondColor ("Second Color", Color) = (0,0,0,0)
        _PowValue ("Pow Value", Range(0,10)) = 5.5
}
```

Next, we need to declare these new properties in our shader:

```
float4 _FirstColor;
float4 _SecondColor;
float  _PowValue;
```

After the announcement of the properties in the shader code, we can use these variables as shown here:

```
void surf (Input IN, inout SurfaceOutput surface)
{
        float4 c = pow(_FirstColor + _SecondColor, _PowValue);
        surface.Albedo = c.rgb;
        surface.Alpha = c.a;
}
```

As a result, you should have a shader as shown here:

```
Shader "PacktPub/YourDiffuseLighting"
{
        Properties
        {
                _FirstColor ("First Color", Color) = (1,1,1,1)
                _SecondColor ("Second Color", Color) = (0,0,0,0)
                _PowValue ("Pow Value", Range(0,10)) = 3.5
        }

        SubShader
        {
                Tags {"RenderType" = "Opaque"}
                LOD 200

                CGPROGRAM
                #pragma surface surf Lambert

                float4 _FirstColor;
                float4 _SecondColor;
                float  _PowValue;
                float4 c;
                struct Input
                {
                        float2 uv_MainTex;
                };

                void surf (Input IN, inout SurfaceOutput surface)
                {
                        c = pow(_FirstColor + _SecondColor, _
    PowValue);
                        surface.Albedo = c.rgb;
                        surface.Alpha = c.a;
                }
                ENDCG
        }

        FallBack "Diffuse"
}
```

It's time to create your own diffuse lighting model. In most cases, no built-in lighting is well-suited for specific tasks in the game or application. Specific optimization problems require unique solutions. In order to override the built-in lighting functions you need to register, the `SubShader` block in the next line of code:

```
#pragma surface surf YourName
```

Now, we can describe our custom lighting function as shown in the following example:

```
inline float4 LightingYourName
(SurfaceOutput surface, float3 lightDirection, float attenuation)
{
        float delta = max(0, dot(surface.Normal, lightDirection));
        c.rgb = (surface.Albedo * _LightColor0.rgb) *
                                            (delta * attenuation *
2);
        c.a = surface.Alpha;
        return c;
}
```

Now, let's systematically look at the fundamental elements. The `#pragma` directive specifies the name of the function to be used for lighting. We used the built-in feature called **Lambert** defined in the `Lighting.cginc` file, and now we specified the name of our function for future use. In establishing this lighting function, it is necessary to remember that the name of the function will eventually be formed using the first word of `Lighting + <Your Function Name>`, that is, for example if you decide to call a function `SunShine`, then the name of your lighting function will be `LightingSunShine`. There are three options to create your custom lighting functions that differ from each other by their input parameters as shown here:

- `float4 Lighting<YourName> (SurfaceOutput surface, float3 lightDirection, float attenuation) {}`: You should use this function for forward rendering, when you don't need the view direction value

- `float4 Lighting<YourName> (SurfaceOutput surface, float3 lightDirection, float3 viewDirection, float attenuation) {}`: You should use this function for forward rendering, when you need the view direction value

- `float4 Lighting<YourName>_PrePass (SurfaceOutput surface, float4 light) {}`: You should use this function for deferred rendering

Ultimately, you should get a shader as follows:

```
Shader "PacktPub/YourLightingModel"
{
        Properties
        {
                _FirstColor ("First Color", Color) = (1,1,1,1)
                _SecondColor ("Second Color", Color) = (0,0,0,0)
                _PowValue ("Pow Value", Range(0,10)) = 3.5
        }

        SubShader
        {
                Tags {"RenderType" = "Opaque"}
                LOD 200

                CGPROGRAM
                #pragma surface surf YourName

                float4 _FirstColor;
                float4 _SecondColor;
                float  _PowValue;
                float4 c;

                struct Input
                {
                        float2 uv_MainTex;
                };

                inline float4 LightingYourName (
                        SurfaceOutput surface,
                        float3 lightDirection,
                        float attenuation
                ){
                        float delta = max(0, dot(surface.Normal,
lightDirection));
                        c.rgb = (surface.Albedo * _LightColor0.rgb) *
                                        (delta * attenuation *
2);

                        c.a = surface.Alpha;
                        return c;
                }
```

```
                        void surf (Input IN, inout SurfaceOutput surface)
                        {
                                c = pow(_FirstColor + _SecondColor, _
        PowValue);

                                surface.Albedo = c.rgb;
                                surface.Alpha = c.a;
                        }
                        ENDCG
                }


                FallBack "Diffuse"
        }
```

A basic reflection environment

Next, let's look at other ideas and techniques widely known in professional circles
around the world that are used to write performance friendly shaders with nice
visual effects. The following examples are based on environment reflections on
your surface. A simple source code of this shader is shown here:

```
Shader "PacktPub/BasicReflectionEnvironment"
{
        Properties
        {
                _DiffuseTint ("Diffuse Tint", Color) = (1,1,1,1)
                _MainTex ("Base (RGB)", 2D) = "white" {}
                _CubeMapTexture ("Cube Map Texture", CUBE) = ""{}
                _ReflectionCount ("Reflection Count", Range(0.01, 1))
        = 0.17
        }


        SubShader
        {
                Tags {"RenderType"="Opaque"}
                LOD 200

                CGPROGRAM
                #pragma surface surf Lambert

                sampler2D _MainTex;
                samplerCUBE _CubeMapTexture;

                float4 _DiffuseTint;
```

```
                float _ReflectionCount;

                float4 c;

                struct Input
                {
                        float2 uv_MainTex;
                        float3 worldRefl;
                };

                void surf (Input IN, inout SurfaceOutput surface)
                {
                        c = tex2D (_MainTex, IN.uv_MainTex) * _
DiffuseTint;
                        surface.Emission = texCUBE(_CubeMapTexture,
IN.worldRefl).rgb * _ReflectionCount;
                        surface.Albedo = c.rgb;
                        surface.Alpha = c.a;
                }
                ENDCG
        }

        FallBack "Diffuse"
}
```

Masked texture reflection

The next new shader implements a new technique that uses a texture to mask the reflection of your environment, and it is shown here:

```
Shader "PacktPub/MaskedTextureReflection"
{
        Properties
        {
                _DiffuseTint ("Diffuse Tint", Color) = (1,1,1,1)
                _MainTex ("Base (RGB)", 2D) = "white" {}
                _ReflectionCount ("Reflection Count", Range(0, 1)) = 1
                _CubeMapTexture ("Cube Map Texture", CUBE) = ""{}
                _MaskedTextureReflection ("Masked Texture Reflection",
2D) = ""{}
        }

        SubShader
```

```
        {
                Tags {"RenderType"="Opaque"}
                LOD 200

                CGPROGRAM
                #pragma surface surf Lambert

                sampler2D _MainTex;
                sampler2D _MaskedTextureReflection;

                samplerCUBE _CubeMapTexture;

                float4 _DiffuseTint;
                float _ReflectionCount;

                float4 c;

                struct Input
                {
                        float2 uv_MainTex;
                        float3 worldRefl;
                };

                void surf (Input IN, inout SurfaceOutput surface)
                {
                        c = tex2D (_MainTex, IN.uv_MainTex);
                        float3 reflectionTexCube = texCUBE(_
CubeMapTexture, IN.worldRefl).rgb;
                        float4 reflectionMaskTexel = tex2D(_
MaskedTextureReflection, IN.uv_MainTex);

                        surface.Albedo = c.rgb * _DiffuseTint;
                        surface.Emission = (reflectionTexCube *
reflectionMaskTexel.r) * _ReflectionCount;
                        surface.Alpha = c.a;
                }
                ENDCG
        }

        FallBack "Diffuse"
}
```

Lighting model techniques

Let's consider a variety of techniques and methods of implementation of the lighting model, which, like the previous shaders, are used worldwide in the game industry as well as in movies and cartoons.

The Lit sphere model

To begin, we would like to consider the LitSphere lighting model. The idea is very simple and straightforward—we should just use a 2D texture in order to completely bake our light. Alternatively, it is necessary to take into account and not forget that this technique is static and does not change the lighting up until the texture used for baking light is changed. This technique gives very high-quality lighting and is optimized enough, but it is not dynamic. In other words, it does not depend on the angle or distance from the camera or from the viewer that can be changed in real time, because this technique does not depend on the lighting in the scene. Let's explore this shader as follows:

```
Shader "PacktPub/LitSphere"
{
        Properties
        {
                _DiffuseTint ("Diffuse Tint", Color) = (1,1,1,1)
                _MainTex ("Base (RGB)", 2D) = "white" {}
                _NormalMapTexture ("Normal Map Texture", 2D) = "bump"
{}
        }

        SubShader
        {
                Tags {"RenderType"="Opaque"}
                LOD 200

                CGPROGRAM
                #pragma surface surf YourUnlit vertex:vert

                sampler2D _MainTex;
                sampler2D _NormalMapTexture;
                float4 _DiffuseTint;

                float4 c;
                float2 uv;
```

```
            inline float4 LightingYourUnlit (SurfaceOutput
surface, float3 lightDirection, float attenuation)
            {
                    c.rgb = float4(1,1,1,1) * surface.Albedo;
                    c.a = surface.Alpha;

                    return c;
            }

            struct Input
            {
                    float2 uv_MainTex;
                    float2 uv_NormalMapTexture;

                    float3 tangentOne;
                    float3 tangentTwo;
            };

void vert (inout appdata_full v, out Input inputData)
{
        UNITY_INITIALIZE_OUTPUT(Input, inputData);

        TANGENT_SPACE_ROTATION;

        inputData.tangentOne = mul(rotation, UNITY_MATRIX_IT_MV[0].
xyz);
        inputData.tangentTwo = mul(rotation, UNITY_MATRIX_IT_MV[1].
xyz);
}

            void surf (Input IN, inout SurfaceOutput surface)
            {
                    surface.Normal = UnpackNormal(tex2D(_
NormalMapTexture, IN.uv_NormalMapTexture)).rgb;

                    uv.x = dot(IN.tangentOne, surface.Normal);
                    uv.y = dot(IN.tangentTwo, surface.Normal);

                    c = tex2D (_MainTex, uv * 0.5 + 0.5);
                    surface.Albedo = c.rgb * _DiffuseTint;
                    surface.Alpha = c.a;
            }
            ENDCG
```

```
    }

        FallBack "Diffuse"
}
```

There are many different techniques and approaches to create lighting models and other visual effects; we cannot put all the ideas and techniques into this book, as it is beyond the scope of this book. You can also implement your own new ideas and techniques; it depends on your imagination. The previous examples of different approaches for writing shaders are widely used by developers around the world to create high-quality rendering in real time, and for optimization. Also, you can write shaders that work with the model vertices, so you can very simply create shader that will play the waves animation from the primitive plane.

Realistic rendering

Marcos Fajardo of *Solid Angle* — the company behind the renderer *Arnold* — noted that more and more production studios in the world have either already come to that, or are in the process of transition with the following quote:

> *"The process is going on in the entire industry, and it's something. I've been working with this for the past ten years or so, and I'm really glad to see that this is happening at last."*

Fajardo can be called one of the greatest defenders and activists of global change in the industry. Solid Angle is really at the forefront of the large-scale movement to the path-traced GI with physically-plausible materials and lighting for production decisions (that is, when the budget is smaller and the time frame is more compressed).

The basis of the popularity of the "honest" method is the desire to "catch two rabbits at once," simplify the lives of artists around the world and to achieve a more realistic picture.

With some older technology chains, artists can get the scene with a few hundred light sources (where each source is needed to fulfill its role, one for the highlight of a material and the second for the specular reflection on this material, the third and fourth for the glare and reflections on the second material, plus ten to simulate global illumination, and so on), with very complex shaders written by the developer with C++, the code was full of tricks and tweaks. Lighting designers often just sit, turning on and off lights, one after the other — it is easy to understand why some of them are so necessary.

Most companies do not count the fact that the introduction of natural light sources and materials make a quick render itself, but there is an expectation that this will greatly facilitate the work of the artist. An hour of work which, to be honest, is a few tens of times more expensive than an hour rendering.

Global illumination

During the, *Game Developers Conference* in 2014, which started on March 17 in San Francisco, the company Unity Technologies introduced the fifth generation of its popular game engine Unity. One of the most important features that distinguishes it from the previous edition is a new system of global illumination in real time—Enlighten—implemented with the participation of experts from the British company, *Geomerics*.

Unity 5 is presented with support from WebGL standard web module optimization `asm.js`, physics engine NVIDIA PhysX 3.3, the system of creation and animation vegetation—SpeedTree, an advanced shader system, the preview function light maps in real time, and a cross-processed audio advertising network in Unity Cloud that facilitates the promotion of mobile games. In addition, the fifth version of the engine will be able to run in a 64-bit environment that significantly simplifies the workflow through a new multi-threaded scheduler, and provides you with the opportunity to make changes in real time and improve the system to create game resources (assets list) and an intuitive interface.

The latest version of this toolkit, traditionally the best-selling in small teams of developers, is designed including large companies. This makes it a competitor to other high-tech new generation of game engines such as CryEngine and Unreal Engine 4. Along with the fourth generation engine from Epic Games, it has recently signed another agreement with Mozilla, where Unity 5 can be used by developers of three-dimensional and two-dimensional games for browsers and mobile devices.

In Unity 5, developers will be able to view the coverage maps in real time using the ray tracing PowerVR from the company *Imagination Technologies*. This technique reduces the processing time, which gives a very good performance. Developers will be able to create a variety of materials from the real world with the new shader system in Unity 5.

Global illumination is the name of a series of algorithms used in three-dimensional graphics for more realistic simulation of light. These algorithms take into account not only the direct light from the source (direct illumination), but also reflected light from various surfaces (indirect illumination).

In theory, *reflection*, *refraction*, and *shadows* are examples of global illumination, because for them it is necessary to consider the effect of the simulation of one object to the other (in contrast to the case when the object is exposed to direct light). In practice, however, the simulation of diffuse reflection or caustics is called global illumination.

Images obtained by the application of global illumination algorithms often appear more realistic than those in the rendering process that apply only direct illumination algorithms. However, to calculate global illumination requires much more time.

The following figure was processed only by direct illumination algorithms:

The following figure was processed by global illumination algorithms:

Practicing in shader optimization

Now, it is time to discuss how we can optimize our shaders. Alternatively, it is time to consider the other methods, such as optimizing built-in data types, which can significantly reduce the overhead of the Unity shaders' memory. We consider Unity shaders optimization for all supported platforms without any exclusions.

Very often, you will need to optimize shaders to achieve the same visual effect, but with a smaller number of textures for example. Primarily, when optimizing shader code, we would like to direct your attention to the types of variables. If you are willing to sacrifice the accuracy of calculations in order to decrease the quality to improve performance, then you should use the `half` or `fixed` variable types instead of `float`. As an example, you can use a `half` type of variable everywhere in your shader code:

```
inline half4 LightingCarVehicle (SurfaceOutput surface, half3
lightDirection, half3 viewDirection, half attenuation)
```

You can also replace `float` with `half` in the following statement:

```
inline float4 LightingCarVehicle (SurfaceOutput surface, float3
lightDirection, float3 viewDirection, float attenuation)
```

- `float`: These variables have 32 bits precision
- `half`: These variables have 16 bits precision
- `fixed`: These variables have 11 bits precision

For example, let's optimize our previous shader code `CarVehicle.shader` as follows:

```
Shader "PacktPub/OptimizedCarVehicle"
{
        Properties
        {
                _DiffuseTint ("Diffuse Tint", Color) = (1,1,1,1)
                _MainTex ("Base (RGB)", 2D) = "white" {}
                _DiffuseIntensity ("Diffuse Intensity", Range(0.01,
17)) = 7.7
                _SpecularColor ("Specular Color", Color) = (1,1,1,1)
                _SpecularIntensity ("Specular Intensity", Range(0.01,
50)) = 17
                _ReflectionCubeMap ("Reflection Cube Map", CUBE) = ""
{}
                _BRDFTexture ("BRDF Texture", 2D) = "white" {}
                _ReflectionIntensity ("Reflection Intensity",
Range(0.01, 11.0)) = 5.0
```

```
                     _ReflectionCount ("Reflection Count", Range(0.01,
     1.0)) = 0.17
                     _FalloffSpread ("Falloff Spread", Range(0.01, 17)) =
     5.3
             }

             SubShader
             {
                     Tags {"RenderType"="Opaque"}
                     LOD 200

                     CGPROGRAM
                     #pragma surface surf CarVehicle

                     samplerCUBE _ReflectionCubeMap;

                     sampler2D _MainTex;
                     sampler2D _BRDFTexture;

                     fixed _SpecularIntensity;
                     fixed _DiffuseIntensity;
                     fixed _FalloffSpread;
                     fixed _ReflectionCount;
                     fixed _ReflectionIntensity;

                     fixed4 _DiffuseTint;
                     fixed4 _SpecularColor;

                     fixed4 c;
                     fixed3 halfVec;
                     fixed falloff;
                     fixed delta;
                     fixed halfVecDotSurfaceNormal;
                     fixed s;

                     inline fixed4 LightingCarVehicle (SurfaceOutput
     surface, fixed3 lightDirection, fixed3 viewDirection, fixed
     attenuation)
                     {
                             halfVec = normalize (lightDirection +
     viewDirection);
                             delta = max (0, dot (surface.Normal,
     lightDirection));

                             halfVecDotSurfaceNormal = 1 - dot(halfVec,
     normalize(surface.Normal));
```

```
                halfVecDotSurfaceNormal = pow(clamp(halfVecDot
SurfaceNormal, 0.0, 1.0), _DiffuseIntensity);
                c = tex2D(_BRDFTexture, fixed2(delta, 1 -
halfVecDotSurfaceNormal));

                s = pow (max (0, dot (surface.Normal,
halfVec)), surface.Specular * _SpecularIntensity) * surface.Gloss;

                c.rgb = (surface.Albedo * _LightColor0.rgb *
c.rgb + _LightColor0.rgb * _SpecularColor.rgb * s)* (attenuation * 2);
                c.a = surface.Alpha + _LightColor0.a * _
SpecularColor.a * s * attenuation;

                return c;
        }

        struct Input
        {
                fixed2 uv_MainTex;

                fixed3 worldRefl;

                fixed3 viewDir;
        };

        void surf (Input IN, inout SurfaceOutput surface)
        {
                c = tex2D (_MainTex, IN.uv_MainTex);

                falloff = pow(saturate(1 - dot(normalize(IN.
viewDir), surface.Normal)), _FalloffSpread);

                surface.Albedo = c.rgb * _DiffuseTint;
                surface.Emission = pow((texCUBE(_
ReflectionCubeMap, IN.worldRefl).rgb * falloff), _ReflectionIntensity)
* _ReflectionCount;
                surface.Specular = c.r;
                surface.Gloss = 1.0;
                surface.Alpha = c.a;
        }
        ENDCG
    }

    FallBack "Diffuse"
}
```

To understand how to develop shaders faster and better, you must understand that this can only be achieved by full optimization, using a variety of techniques and approaches. Let's divide our shader optimization process into the following three categories:

- Optimization of memory used for variables
- Optimization of the number and size of used textures
- Optimization of computational algorithms

All the mentioned points have been already discussed. Some of the ideas of how to optimize your shaders have been already considered earlier in this chapter, and we will show some more interesting approaches and techniques. We hope that most of the methods, techniques, approaches, and ideas of this book will greatly help you in production to achieve the desired quality and performance at the same time.

Also, while optimizing shaders you need to remember and know that the code should be as small as possible. This means that nothing unnecessary should be in your code. Many of the ideas described in the previous chapters, especially in the preceding fourth chapter, which talked about code optimization in C# and JavaScript, are well suited to optimize your shader code. Also, we want you to note that the frequency of execution of your shader code greatly affects the performance. Very often, shader developers use very good techniques in order to optimize their shaders. They prefer using vertex shaders instead of pixel shaders; this will greatly improve your performance in most cases, as there are significantly more pixels than vertices. Hence, processing the pixel shader frequency of the execution code will be much greater than for vertices.

Let's also consider the directives that may well optimize your shaders:

- `approxview`: This approximation is good enough in many cases. You should use this directive when you need to get the normalized view direction per vertex instead of per pixel.
- `halfasview`: This will be computed and normalized per vertex halfway between view and light directions (half vector), and the lighting function will receive a `half` vector as a parameter instead of view vector.
- `noforwardadd`: In the event of rendering shaders in one pass, even with multiple lights, and in the event of making your shader smaller, this directive is the best choice for you. The shader will support only one directional light in forward rendering. The rest of the lights can still have an effect like vertex lights or spherical harmonics. The rest of the lights can be used for spherical harmonics or for vertex light effects.
- `exclude_path:prepass`: The shader that uses this directive will not accept any custom lighting from the deferred renderer.

- noambient: This directive should be used in the event of deactivating spherical harmonics and ambient lights on your shader. This can slightly enhance your performance.

- nolightmap: This directive disables Unity's internal light mapping system. In other words, it does not perform a light mapping check.

Alpha testing on mobile devices is very expensive, so you should use transparent shaders on mobile devices very accurately. You must use the alpha testing only when necessary. For example, let's cover optimized shader as follows:

```
Shader "PacktPub/OptimizedShaderExample"
{
        Properties
        {
                _MainTex ("Base (RGB)", 2D) = "white" {}
                _SpecularWidth ("Specular Width", Range(0.01, 1)) =
0.5
                _NormalMapTexture ("Normal Map Texture", 2D) =
"bump"{}
        }

        SubShader
        {
                Tags {"RenderType"="Opaque"}
                LOD 200

                CGPROGRAM
                #pragma surface surf OptimizedBlinnPhong exclude_
path:prepass nolightmap noforwardadd halfasview

                sampler2D _MainTex;
                sampler2D _NormalMapTexture;
                half _SpecularWidth;

                half4 c;
                half d;
                half s;

                struct Input
                {
                        half2 uv_MainTex;
                };

                inline half4 LightingOptimizedBlinnPhong
        (SurfaceOutput surface, half3 lightDir, half3 halfDir, half atten)
```

```
                 {
                        d = max(0, dot(surface.Normal, lightDir));
                        s = pow(max(0, dot(surface.Normal, halfDir)),
surface.Specular * 128) * surface.Gloss;

                        c.rgb = (surface.Albedo * _LightColor0.rgb * d
+ _LightColor0.rgb * s) * (atten * 2);
                        c.a = 0.0;

                        return c;
                 }

                 void surf (Input IN, inout SurfaceOutput surface)
                 {
                        c = tex2D(_MainTex, IN.uv_MainTex);

                        surface.Albedo = c.rgb;
                        surface.Gloss = c.a;
                        surface.Alpha = 0.0;
                        surface.Specular = _SpecularWidth;
                        surface.Normal = UnpackNormal(tex2D(_
NormalMapTexture, IN.uv_MainTex)).rgb;
                 }
                 ENDCG
         }

         FallBack "Diffuse"
}
```

Best case practice

After our journey into the world of diverse ideas, approaches, and techniques in
the field of lighting calculations, let's look at the best practice in order to maintain
many different shaders easily. Let's consider the possibility of reusability of our
shader code; for example, various lighting functions in Unity. In order to avoid
writing the same code for the same lighting function each time for a new shader,
it's best to write the code lighting function once and then just use this in any shader
if necessary, as programmers use different frameworks and libraries. This practice
will help you create a framework for your shaders, which will greatly facilitate easy
development and effortless shader maintenance. In the previous examples, we used
built-in CgIncludes files such as Lambert and BlinnPhong lighting functions. Unity
created these lighting models for us. Unity helps us reduce our efforts while writing
performance-friendly and nice quality shaders.

You can view the code embedded in Unity built-in CgIncludes files, which are located inside a directory called CgIncludes. Without these files, it will be much harder to write shaders in Unity. That's why Unity surface shaders are so efficient. Let's create our own CgInclude file as follows:

```
#ifndef YOUR_NAME_INCLUDE
#define YOUR_NAME_INCLUDE


half4 _YourColorVariable;


inline half4 LightingOptimizedLambert (SurfaceOutput surface, half3
lightDirection, half attenuation)
{
        half diffuseValue = max(0, dot(surface.Normal,
lightDirection));
        diffuseValue = (diffuseValue + 0.5) * 0.5;

        half4 tmpColor;
        tmpColor.rgb = surface.Albedo * _LightColor0.rgb *
((diffuseValue * _YourColorVariable.rgb) * attenuation * 2);
        tmpColor.a = surface.Alpha;

        return tmpColor;
}


#endif
```

Now let's consider the code for the next shader as follows, where you can see how to use your CgInclude file with your lighting function, and how you can declare your variable _YourColorVariable:

```
Shader "PacktPub/UsingCgIncludeOptimzedLambert"
{
        Properties
        {
                _YourColorVariable ("Your Color Variable", Color) =
(1,1,1,1)

                _DiffuseTint ("Diffuse Tint", Color) = (1,1,1,1)
                _MainTex ("Base (RGB)", 2D) = "white" {}
                _NormalMapTexture ("Normal Map Texture", 2D) = "bump"
{}
                _CubeMapTexture ("Cube Map Texture", CUBE) = ""{}
                _ReflectionCount ("Reflection Count", Range(0,1)) =
0.17
```

```
        }

    SubShader
    {
            Tags {"RenderType"="Opaque"}
            LOD 200

            CGPROGRAM
            #include "YourCgIncludeOptimizedLambert.cginc"
            #pragma surface surf OptimizedLambert

            samplerCUBE _CubeMapTexture;

            sampler2D _MainTex;
            sampler2D _NormalMapTexture;

            float4 _DiffuseTint;
            float _ReflectionCount;

            float4 c;

            struct Input
            {
                    float2 uv_MainTex;

                    float2 uv_NormalMapTexture;

                    float3 worldRefl;

                    INTERNAL_DATA
            };

            void surf (Input IN, inout SurfaceOutput surface)
            {
                    c = tex2D (_MainTex, IN.uv_MainTex);

                    surface.Normal = UnpackNormal(tex2D(_
NormalMapTexture, IN.uv_NormalMapTexture)).rgb;
                    surface.Emission = texCUBE (_CubeMapTexture,
WorldReflectionVector(IN, surface.Normal)).rgb * _ReflectionCount;
                    surface.Albedo = c.rgb * _DiffuseTint;
                    surface.Alpha = c.a;
```

```
            }
            ENDCG
        }

    FallBack "Diffuse"
}
```

Thus, you can create a proper framework of your own shaders. Also, you can use examples from this chapter and place all your code in the form of `CgIncludes` files. This will greatly help you avoid repetitions in the code, which will greatly simplify the development of shaders and also facilitate their optimization.

Summary

In this chapter, you learned a lot about writing shaders and their optimization. We started with a simple shader code and examined the fundamental elements in Unity surface shaders. Next, we wrote our custom diffuse lighting model. Also, we examined global illumination. We explored the various optimization techniques by changing the shader variable types, as well as by writing specific directives. Towards the end of this chapter, we covered the best case practice while developing shaders using `CgIncludes` files, and learned how to use its code.

The next chapter will cover legacy and Mecanim animation systems in Unity 5. You will also develop a simple custom sprite animation system and explore how to import, set up, and play audio files inside your scripts in Unity 5. At the end of the next chapter, you will explore the physics and particle systems in Unity 5.

4
Animation, Audio, Physics, and Particle Systems in Unity 5

In this chapter, you will learn new Mecanim animation features and awesome new audio features in Unity 5. At the end of this chapter, you will explore physics and particle systems in Unity 5.

The topics that will be covered in the chapter are as follows:

- New Mecanim animation features in Unity 5
- New audio features in Unity 5
- Physics and particle system effects in Unity 5

New Mecanim animation features in Unity 5

Unity 5 contains some new awesome possibilities for the Mecanim animation system. Let's look at the new shiny features known in Unity 5.

State machine behavior

Now, you can inherit your classes from `StateMachineBehaviour` in order to be able to attach them to your Mecanim animation states. This class has the following very important callbacks:

- `OnStateEnter`
- `OnStateUpdate`
- `OnStateExit`
- `OnStateMove`
- `OnStateIK`

The `StateMachineBehaviour` scripts behave like `MonoBehaviour` scripts, which you can attach on as many objects as you wish; the same is true for `StateMachineBehaviour`. You can use this solution with or without any animation at all.

State machine transition

Unity 5 introduced a new awesome feature for Mecanim animation systems known as state machine transitions in order to construct a higher abstraction level. In addition, entry and exit nodes were created. By these two additional nodes to `StateMachine`, you can now branch your start or finish state depending on your special conditions and requirements.

> These mixes of transitions are possible: `StateMachine` | `StateMachine`, `State` | `StateMachine`, `State` | `State`.

In addition, you also can reorder your layers or parameters. This is the new UI that allows it by a very simple and useful drag-n-drop method.

Asset creation API

One more awesome possibility in Unity 5 was introduced using scripts in Unity Editor in order to programmatically create assets, such as layers, controllers, states, `StateMachine`, and blend trees. You can use different solutions with a high-level API provided by Unity engine maintenance and a low-level API, where you should manage all your assets manually. You can find more about both API versions on Unity documentation pages.

Direct blend tree

Another new feature that was introduced with the new BlendTree type is known as direct. It provides direct mapping and animator parameters to the weight of BlendTree children.

 Possibilities with Unity 5 have been enhanced with two useful features for Mecanim animation system:
- Camera can scale, orbit, and pan
- You can access your parameters in runtime

Programmatically creating assets by Unity 5 API

The following code snippets are self-explanatory, pretty simple, and straightforward. I list them just as a very useful reminder.

Creating the controller

To create a controller you can use the following code:

```
var animatorController = UnityEditor.Animations.AnimatorController.
CreateAnimatorControllerAtPath ("Assets/Your/Folder/Name/state_
machine_transitions.controller");
```

Adding parameters

To add parameters to the controller, you can use this code:

```
animatorController.AddParameter("Parameter1", UnityEditor.Animations.
AnimatorControllerParameterType.Trigger);
animatorController.AddParameter("Parameter2", UnityEditor.Animations.
AnimatorControllerParameterType.Trigger);
animatorController.AddParameter("Parameter3", UnityEditor.Animations.
AnimatorControllerParameterType.Trigger);
```

Adding state machines

To add state machines, you can use the following code:

```
var sm1 = animatorController.layers[0].stateMachine;
var sm2 = sm1.AddStateMachine("sm2");
var sm3 = sm1.AddStateMachine("sm3");
```

Adding states

To add states, you can use the code given here:

```
var s1 = sm2.AddState("s1");
var s2 = sm3.AddState("s2");
var s3 = sm3.AddState("s3");
```

Adding transitions

To add transitions, you can use the following code:

```
var exitTransition = s1.AddExitTransition();
exitTransition.AddCondition(UnityEditor.Animations.
AnimatorConditionMode.If, 0, "Parameter1");
exitTransition.duration = 0;

var transition1 = sm2.AddAnyStateTransition(s1);
transition.AddCondition(UnityEditor.Animations.AnimatorConditionMode.
If, 0, "Parameter2");
transition.duration = 0;

var transition2 = sm3.AddEntryTransition(s2);
transition2.AddCondition(UnityEditor.Animations.AnimatorConditionMode.
If, 0, "Parameter3");
sm3.AddEntryTransition(s3);
sm3.defaultState = s2;

var exitTransition = s3.AddExitTransition();
exitTransition.AddCondition(UnityEditor.Animations.
AnimatorConditionMode.If, 0, "Parameter3");
exitTransition.duration = 0;

var smt = rootStateMachine.AddStateMachineTransition(sm2, sm3);
smt.AddCondition(UnityEditor.Animations.AnimatorConditionMode.If, 0,
"Parameter2");
sm2.AddStateMachineTransition(sm1, sm3);
```

Going deeper into new audio features

Let's start with new amazing Audio Mixer possibilities.

Now, you can do true submixing of audio in Unity 5.

In the following figure, you can see a very simple example with different sound categories required in a game:

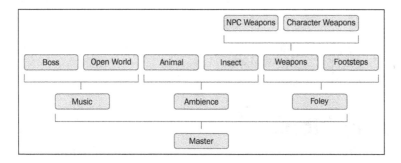

Now in Unity 5, you can mix different sound collections within categories and tune up volume control and effects only once in a single place so that you can save a lot of time and effort. This new awesome audio feature in Unity 5 allows you to create a fantastic mood and atmosphere for your game.

Each Audio Mixer can have a hierarchy of AudioGroups:

The Audio Mixer can not only do a lot of useful things, but also mix different sound groups in one place. Different audio effects are applied sequentially in each AudioGroup.

Now you're getting closer to the amazing, awesome, and shiny new features in Unity 5 for audio system! A callback script OnAudioFilterRead, which made possible the processing of samples directly into their scripts, previously was handled exclusively by the code.

Unity now also supports custom plugins to create different effects. With these innovations, Unity 5 for audio system now has its own applications synthesizer, which has become much easier and more flexible than possible.

Mood transitions

As mentioned earlier, the mood of the game can be controlled with a mix of sound. This can be achieved with the involvement of new stems and music or ambient sounds. Another common way to accomplish this is to move the state of the mixture. A very effective way of taking mood where you want to go is by changing the volume section's mixture and transferring it to the different states of effect parameters.

Inside, everything is the Audio Mixer's ability to identify pictures. Pictures capture the status of all parameters in Audio Mixer. Everything from investigative wet levels to AudioGroup tone levels can be captured and moved between the various parameters.

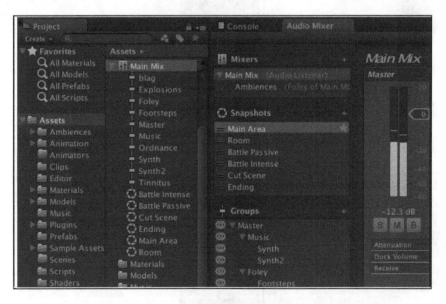

You can even create a complex mixture of states between a whole bunch of pictures in your game, creating all kinds of possibilities and goals.

Imagine installing all these things without having to write a line of code to the script.

Physics and particle system effects in Unity 5

Physics for 2D and 3D in Unity are very similar, because they use the same concepts like Ias rigidbodies, joints, and colliders. However, Box2D has more features than Unity's 2D physics engine. It is not a problem to mix 2D and 3D physics engines (built-in, custom, third-party) in Unity. So, Unity provides an easy development way for your innovative games and applications.

If you need to develop some real-life physics in your project, then you should not write your own library, framework, or engine, except specific requirements. However, you should try existing physics engines, libraries, or frameworks with many features already made.

Let's start our introduction into Unity's built-in physics engine. In the case that you need to set your object under Unity's built-in physics management, you just need to attach the Rigidbody component to this object. After that, your object can collide with other entities in its world and gravity will have an affect on it. In other words, Rigidbody will be simulated physically. In your scripts, you can move any of your Rigidbodies by adding vector forces to them.

It is not recommended to move the Transform component of a non-kinematic Rigidbody, because it will not collide correctly with other items. Instead, you can apply forces and torque to your Rigidbody.

A Rigidbody can be used also to develop cars with wheel colliders and with some of your scripts to apply forces to it. Furthermore, a Rigidbody is used not *only* for vehicles, but also you can use it for any other physics issues such as airplanes, robots with various scripts for applying forces, and with joints.

The most useful way to utilize a Rigidbody is to use it in collaboration with some primitive colliders (built-in in Unity) such as BoxCollider and SphereCollider. Next, we will show you two things to remember about Rigidbody:

- In your object's hierarchy, you must never have a child and its parent with the Rigidbody component together at the same time
- It is not recommended to scale Rigidbody's parent object

One of the most important and fundamental components of physics in Unity is a Rigidbody component. This component activates physics calculations on the attached object. If you need your object to react to collisions(for example, while playing billiards, balls collide with each other and scatter in different directions) then you must also attach a `Collider` component on your GameObject. If you have attached a Rigidbody component to your object, then your object will move through the physics engine, and I recommend that you do not move your object by changing its position or rotation in the `Transform` component. If you need some way to move your object, you should apply the various forces acting on the object so that the Unity physics engine assumes all obligations for the calculation of collisions and moving dynamic objects. Also, in some situations, there is a need for a Rigidbody component, but your object must be moved only by changing its position or rotation properties in the `Transform` component. It is sometimes necessary to use components without Rigidbody calculating collisions of the object and its motion physics. That is, your object will move by your script or, for example, by running your animation. In order to solve this problem, you should just activate its `IsKinematic` property. Sometimes, it is required to use a combination of these two modes when `IsKinematic` is turned on and when it is turned off. You can create a symbiosis of these two modes, changing the `IsKinematic` parameter directly in your code or in your animation.

Changing the `IsKinematic` property very often from your code or from your animation can be the cause of overhead in your performance. Therefore, you should use it very carefully and only when you really need it.

A kinematic Rigidbody object is defined by the `IsKinematic` toggle option. If a Rigidbody is `Kinematic`, this object will not be affected by collisions, gravity, or forces.

There is a Rigidbody component for 3D physics engine and an analogous Rigidbody2D for 2D physics engine.

A kinematic Rigidbody can interact with other non-kinematic Rigidbodies. In the event of using kinematic Rigidbodies, you should translate their positions and rotation values of the `Transform` component by your scripts or animations. When there is a collision between Kinematic and non-kinematic Rigidbodies, then the Kinematic object will properly wake up non-kinematic Rigidbody. Furthermore, the first Rigidbody will apply friction to the second Rigidbody if the second object is on top of the first object.

Let's list some possible usage examples of kinematic Rigidbodies:

- There are situations when you need your objects to be under physics management, but sometimes to be controlled explicitly from your scripts or animations. As an example, you can attach Rigidbodies to the bones of your animated personage and connect them with joints in order to utilize your entity as a ragdoll. If you are controlling your character by Unity's animation system, you should enable the IsKinematic checkbox. Sometimes you may require your hero to be affected by Unity's built-in physics engine if you are hitting the hero. In this case you should disable the IsKinematic checkbox.

- If you need a moving item that can push different items, yet not by itself. In case you have a moving platform and you need to place some Rigidbody objects on top, you ought to enable the IsKinematic checkbox rather than simply attaching a collider without a Rigidbody.

- You may need to enable the IsKinematic property of your Rigidbody object that is animated and has a genuine Rigidbody follower by utilizing one of the accessible joints.

Earlier, I mentioned the collider, but now is the time to discuss this component in more detail. In the case of Unity, the physics engine can calculate collisions. You must specify geometric shapes for your object by attaching the Collider component. In most cases, the collider does not have to be the same shape as your mesh with many polygons. Therefore, it is desirable to use simple colliders, which will significantly improve your performance, otherwise with more complex geometric shapes you risk significantly increasing the computing time for physics collisions. Simple colliders in Unity are known as primitive colliders: BoxCollider, BoxCollider2D, SphereCollider, CircleCollider2D, and CapsuleCollider. Also, no one forbids you to combine different primitive colliders to create a more realistic geometric shape that the physics engine can handle very fast compared to MeshCollider. Therefore, to accelerate your performance, you should use primitive colliders wherever possible. You can also hang on to the child objects of different primitive colliders, which will change its position and rotation, depending on the parent Transform component. The Rigidbody component must be attached only to the GameObject root in the hierarchy of your entity.

Unity provides a `MeshCollider` component for 3D physics and a `PolygonCollider2D` component for 2D physics. The `MeshCollider` component will use your object's mesh for its geometric shape. In `PolygonCollider2D`, you can edit directly in Unity and create any 2D geometry for your 2D physical computations. In order to react in collisions between different mesh colliders, you must enable a `Convex` property. You will certainly sacrifice performance for more accurate physics calculations, but if you have the right balance between quality and performance, then you can achieve good performance only through a proper approach.

Objects are static when they have a `Collider` component without a Rigidbody component. Therefore, you should not move or rotate them by changing properties in their `Transform` component, because it will leave a heavy imprint on your performance as a physics engine should recalculate many polygons of various objects for right collisions and ray casts. Dynamic objects are those that have a Rigidbody component. Static objects (attached with the `Collider` component and without Rigidbody components) can interact with dynamic objects (attached with `Collider` and Rigidbody components). Furthermore, static objects will not be moved by collisions like dynamic objects.

Also, Rigidbodies can sleep in order to increase performance. Unity provides the ability to control sleep in a Rigidbodies component directly in the code using following functions:

- `Rigidbody.IsSleeping()`
- `Rigidbody.Sleep()`
- `Rigidbody.WakeUp()`

There are two variables characterized in the physics manager. You can open physics manager right from Unity menu here: **Edit | Project Settings | Physics**:

- `Rigidbody.sleepVelocity`: The default value is `0.14`. This indicates lower limitations for linear velocity (from zero to infinity) below which objects will sleep.
- `Rigidbody.sleepAngularVelocity`: The default value is `0.14`. This indicates lower limitations for angular velocity (from zero to infinity) below which objects will sleep.

Rigidbodies awaken when:

- An alternate Rigidbody impacts the resting Rigidbody
- An alternate Rigidbody was joined through a joint
- At the point of adjusting a property of the Rigidbody
- At the point of adding force vectors

 A kinematic Rigidbody can wake the other sleeping Rigidbodies while static objects (attached with a `Collider` component and without a Rigidbody component) can't wake your sleeping Rigidbodies.

The PhysX physics engine which is integrated into Unity works well on mobile devices, but mobile devices certainly have far fewer resources than powerful desktops.

Let's look at a few points to optimize the physics engine in Unity:

- First of all, note that you can adjust the `Fixed Timestep` parameter in the time manager in order to reduce costs for the physical execution time updates. If you increase the value, you can increase the quality and accuracy of physics in your game or in your application, but you will lose the time to process. This can greatly reduce your productivity, or in other words, it can increase CPU overhead.
- The maximum allowed timestep indicates how much time will be spent in the worst case for physical treatment.
- The total processing time for physics depends on the awake rigidbodies and colliders in the scene, as well as the level of complexity of the colliders.

Unity provides the ability to use physical materials for setting various properties such as friction and elasticity. For example, a piece of ice in your game may have very low friction or equal to zero (minimum value), while a jumping ball may have a very high friction force or equal to one (maximum value) and also very high elasticity. You should play with the settings of your physical materials for different objects and choose the most suitable solution for you and the best solution for your performance.

Triggers do not require a lot of processing costs by the physics engine and can greatly help in improving your performance. Triggers are useful in situations where, for example, in your game you need to identify areas near all lights that are automatically turned on in the evening or night if the player is in its trigger zone or in other words within the geometric shape of its collider, which you can design as you wish. Unity triggers allow writing the three callbacks, which will be called when your object enters the trigger, while your object is staying in trigger, and when this object leaves the trigger. Thus, you can register any of these functions, the necessary instructions, for example, turn on the flashlight when entering the trigger zone or turn it off when exiting the trigger zone. It is important to know that in Unity, static objects (objects without a Rigidbody component) will not cause your callbacks to get into the zone trigger if your trigger does not contain a Rigidbody component; that is, in other words at least one of these objects must have a Rigidbody component in order to not ignore your callbacks. In the case of two triggers, there should be at least one object attached with a Rigidbody component to your callbacks were not ignored. Remember that when two objects are attached with Rigidbody and Collider components and if at least one of them is the trigger, then the trigger callbacks will be called and not the collision callbacks. I would also like to point out that your callbacks will be called for each object included in the collision or trigger zone. Also, you can directly control whether your collider is a trigger or not by setting the flag isTrigger value to true or false in your code. Of course, you can mix both options in order to obtain the best performance. All collision callbacks will be called only if at least one of two interacted rigidbodies is not kinematic. I suggest you consider code samples for collision callbacks.

The first example callback, which will be called at the start of the collision event:

```
void OnCollisionEnter (Collision collision)
```

The second example callback, which will be called while staying in the collision state:

```
void OnCollisionStay (Collision collision)
```

The third example callback, which will be called at the end of the collision event:

```
void OnCollisionExit (Collision collision)
```

The fourth example callback, which will be called at the start of the collision event is much more optimized because it avoids a collision input parameter and thus avoids extra calculations:

```
void OnCollisionEnter ()
```

The fifth example callback, which will be called while staying in the collision state is much more optimized because it avoids a collision input parameter and thus avoids extra calculations:

```
void OnCollisionStay ()
```

The sixth example callback, which will be called at the end of the collision event is more optimized because it avoids the collision input parameter and thus avoids extra calculations:

```
void OnCollisionExit ()
```

The seventh example callback, which will be called on entering the trigger collider:

```
void OnTriggerEnter (Collider collider)
```

The eighth example callback, which will be called while staying in the trigger collider:

```
void OnTriggerStay (Collider collider)
```

The ninth example callback, which will be called on exiting the trigger collider:

```
void OnTriggerExit (Collider collider)
```

The tenth example callback, which will be called on entering the trigger without the collider input parameter. This callback will be faster than with the collider input parameter:

```
void OnTriggerEnter ()
```

The eleventh example callback, which will be called on the enter trigger without the collider input parameter. This callback will be faster than with the collider input parameter:

```
void OnTriggerStay ()
```

The twelfth example callback, which will be called on exiting the trigger without the collider input parameter. This callback will be faster than with the collider input parameter:

```
void OnTriggerExit ()
```

Now, let's talk about joints. If you need to attach one Rigidbody to another Rigidbody in order to rotate it around a specific point and axis, for example a hinged door, then you should use HingeJoint (for 2D appropriate name is HingeJoint2D). Unity also provides other types of joints; for example, spring joint is suitable in cases when you need to develop a trampoline or something similar. However, I strongly advise you not to use joints everywhere because that can ruin your performance. Use only what is truly necessary and as often as is really required. The most important thing to improve your performance is to remove all unnecessary things.

You can also use a `CharacterController` component for creating a first person game. The `CharacterController` component uses its own physics calculations separately from Rigidbody.

These are very convenient for walking on different surfaces around the y axis without rotation and maintaining the necessary balance in the case of a Rigidbody component. The `CharacterController` component also requires much less processing time compared to a Rigidbody. That's why you should try to use `CharacterController` whenever possible instead of Rigidbody, but try to make as few as possible like everything else. The `CharacterController` component has `CapsuleCollider`, which contains upwards along the y axis. Next, you will research the `CharacterController` properties as shown in the following screenshot:

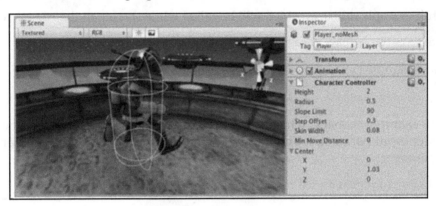

Particle system performance tips and tricks

A particle system uses a large number of small particles or, in other words, a huge amount of graphical objects in order to create different effects such as dust, rain, snow, fire, explosions, smoke, a waterfall, falling leaves, fog, stars, galaxies, fireworks, various magic effects, and so on. Usually a particle system emits a plurality of particles, which have their own life-time, after which they disappear gradually and are re-emitted. There are also different techniques of using a particle system to create fur, hair, grass, where the particles do not disappear, but they live for a very long time.

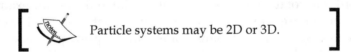

Particle systems may be 2D or 3D.

Mathematically, each particle is represented as a point mass with additional attributes, such as appearance, speed, orientation in space, angular velocity, and so on. In the course of the program, each particle changes its state with a specific formula, common to all particles in the system. For example, the particle may be exposed to gravity, to change its size, color, speed, and so on. After all calculations, the particle will be visualized. A particle can be visualized by point, triangle, sprite, or even a full three-dimensional model.

Currently, there is no uniform implementation of particle systems. In different games and apps, 3D modeling properties, behavior, and appearance of the particles may be fundamentally different.

In most implementations, new particles are emitted by a so-called **emitter**. If the emitter is a point, the new particles will then occur in the same place. Therefore, it is possible to simulate, for example, an explosion—the emitter is its center. An emitter can be a line, segment, or a plane; for example rain or snow particles should occur at high horizontal planes. The emitter may have an arbitrary geometrical object, and in this case, the new particles will emerge on the entire surface thereof.

Throughout the lifetime of the particle, the particle is rarely static. Particles can move, rotate, change color and/or transparency, and may deal with three-dimensional objects. Often, particles set the maximum life span, after which the particle disappears.

In three-dimensional, real-time applications or computer games it is generally considered that the particles do not cast shadows on one another and on the geometry of the environment and they do not absorb and emit light. Without these, the simplified particle system will require more resources; in the case of absorption of light, the particles need to be sorted by distance from the camera, and in the case of each particle shadows have to paint several times.

Legacy versus Shuriken Unity's built-in particle systems

1. Some of Shuriken's module's properties cannot be achieved in your scripts.
2. At the same time, the Legacy particle system's properties can be achieved in your scripts.
3. You can turn `emission` on and off as shown in the following code:

```
public ParticleSystem yourParticleSystemVariable;
void YourMethodName() {
  yourParticleSystemVariable.enableEmission = false;
}
```

In event of emitting particles in explosions, you should use the `Emit` function as shown in the following code example:

```
public ParticleSystem yourParticleSystemVariable;
void YourMethodName() {
  yourParticleSystemVariable.Emit(123); //emits 123 particles
}
```

Instead of activating and deactivating the `emission` property, you can also control the particles in your emitter as shown here:

```
public ParticleSystem yourParticleSystem;
private ParticleSystem.Particle[]
    yourParticlesList = new ParticleSystem.Particle[1750];
void YourMethodName() {
  int len = yourParticleSystem.GetParticles(yourParticlesList);
  for(int i=0; i < len; i++) {
        yourParticlesList[i].color = new Color(0,0,1,0.5f);
    }
  yourParticleSystem.SetParticles(yourParticlesList, len);
}
```

Let's list obtainable properties of the `Particle` class (`http://docs.unity3d.com/ScriptReference/ParticleSystem.Particle.html`):

- `lifetime`
- `startLifetime`
- `position`
- `rotation`
- `color`
- `size`
- `velocity`
- `randomValue`
- `angularVelocity`

Creating the Shuriken particle system in Unity is very simple. You just need to navigate to **GameObject** | **Create Other** | **Particle Systems**. This will create an instance of the Shuriken Particle System for you to play with.

To make a Legacy particle system, you have to make a `void` GameObject or join the Legacy particle system to an accessible GameObject.

Particle system tips

There are numerous things that can kill the frame rate in an up-to-date game, and particles are up close to the highest on the rundown of reasons. A key component is that particles are liable to a great deal of overdraw that is not displayed in your opaque geometry.

The purpose behind the increment in overdraw is that for particles, we have a tendency to have heaps of distinctive primitives (typically quads) that are covered, maybe to copy impacts like flame or smoke. Regularly, every particle primitive is translucent (alpha-mixed), so the z-buffer is not overhauled as pixels are composed and we wind up rendering to pixels at different times. (Interestingly, for hazy geometry, we do keep in touch with the z-buffer, so between a conceivable z-prepass, sorting items front-to-back, progressive z-culling on the GPU, and ordinary profundity testing, the effect is that we have almost no overdraw.)

Overdraw, thus, prompts expanded employments of both `fillrate` (how many pixels the fittings can render to for every second) and `bandwidth` (how much information you can exchange to/from the GPU for every second), both of which may be rare assets.

We concur that particles can result in a considerable measure of issues. Luckily, there are heaps of things that are able to improve the rendering side of a particle system.

- **Use opaque particles**: For example, make smoke effects truly thick so that (some or the sum of) the molecule boards will be obscure, with set pattern alpha. For a few particles, in the same way as shrapnel, rocks, or comparable objects, use lightweight geometry particles rather than sprites with alpha borders.

- **Use wealthier particles**: Put more oomph in a solitary molecule sprite so that you require fewer of them. Use flip book surfaces to make surging in for example fire and smoke, instead of stacking sprites.

- **Top aggregate sum of particles**: Use fittings counters on the graphics card to obtain the number of particle pixels that have been rendered, and quit discharging or drawing particles when passing a certain breaking point (which may be set dynamically).

- **Decrease state changes**: Share shaders between particles. You can get this by, for example, dropping characteristics for far off particles (for example, dropping the normal map at the earliest opportunity).

- **Make particles front-to-back premultiplied-alpha style**: Using premultiplied alpha (which is cooperative), you can mix particles front-to-back rather than the ordinary back-to-front requesting. The thought here is to utilize the front-to-back attracting to fill a stencil buffer when alpha gets (close) strong and at last, quit drawing particles all together (when they generally won't help the visual scene much).

- **Bunch particles together into one molecule entity**: Instead of drawing two covering particles separately, you can structure a solitary (bigger) molecule that incorporates the two particles and performs the mixing of the two particles in a staightforward manner in the shader. This has a tendency to decrease the measure of the frame buffer understands we do, as we just need to mix one molecule.

Summary

This chapter covered new Mecanim animation features in Unity 5. You were introduced to the new awesome audio features in Unity 5. At the end of this chapter, you explored physics and particle systems in Unity 5. In this chapter, you covered many useful details for your performance within Unity built-in physics and particle systems. You explored Rigidbodies and different tips and tricks tied with it. You learned about kinematic and sleeping Rigidbodies, colliders, static colliders, primitive colliders, physics materials, triggers, joints, character controller, interactive cloth, and a lot of other useful physics definitions, details, notes, tips, and tricks. You also learned a lot about the particle system tips and tricks and how to create a simple pooling system for any of your objects.

The next chapter will include an overview about asset bundles in Unity 5. You will also learn how to download new code and data in real-time for Android devices. At the end of this chapter, you will discover the safeness technique of the asset bundles in practice.

5
Asset Bundles in Unity 5 Pro

This chapter will include an overview of Asset Bundles in Unity 5. You will learn how to download new code and data in real time for Android devices. At the end of this chapter, the reader will discover the safeness technique of the asset bundles in practice.

The topics that will be covered in this chapter are as follows:

- Overview of Asset Bundles in Unity 5
- Downloading new code and data in real time for Android devices
- Safeness techniques in practice

An overview of the asset bundles in Unity 5

Asset bundles are **Unity Pro** features only. Two main ideas of the asset bundles are:

- Easily download content in your application
- Uploading new content in your application

Unity allows exporting your assets as files, which are known as asset bundles. Your application can download these compressed files whenever needed. This approach will reduce your final build size by streaming in: prefabs, animations, binary files, textures, audio clips, meshes, and scenes, where asset bundles will be utilized. All other asset types are supported by Unity. For binary files, you should set the extension to .bytes and Unity will recognize these files as TextAsset. To use asset bundles, you just need to create them and upload them to your server. In the Unity Editor, you can build asset bundles from your assets in the scene. In a situation when you need to upload your asset bundles to the server, you can use any data communication protocol; for example, SSH, FTP, FTPS, SFTP, or any other protocol depending on your choice. In real time, your application, which is written in your script, will download the necessary asset bundles for further work with your packed assets in these exported files.

We will cover what you should do to create the AssetBundle file. For this task, you should use the Unity Editor class known as BuildPipeline.

 If you are using any of the Unity Editor classes in your scripts, then you should always remember to keep these scripts in a folder named Editor anywhere in your project within any subdirectory of the Assets folder.

Now, let's create a simple C# script to create AssetBundle. First of all, we should import two Unity namespaces:

```
using UnityEngine;
using UnityEditor;
```

After these lines of code, you should declare your public class; for example, you can create very simple class declaration:

```
public class BuilderAssetBundle {
  // the code you will see below
}
```

In the next step, we will create a static function with its MenuItem for future selection from the Unity Editor menu:

```
[MenuItem("PacktPub/AssetBundles/Build Asset Bundle")]
static void Build() {
  // the code you will see below
}
```

After that, you need to fill in your static function with just a single instruction or line of code or something similar depending on your requirements, and you will see the already finished simple class to build your `AssetBundle`:

```
BuildPipeline.BuildAssetBundle(
  Selection.activeObject,
  Selection.GetFiltered(typeof(Object),SelectionMode.DeepAssets),
  "Assets/Your/Path/To/YourAssetBundle.unity3d",
  BuildAssetBundleOptions.CollectDependencies |
  BuildAssetBundleOptions.CompleteAssets
);
```

This function creates a compressed `Assets/Your/Path/To/YourAssetBundle.unity3d` file with a list of packed assets (any from your project folder) and returns `true` if `AssetBundle` was created successfully or `false` otherwise. The first variable in this function, `Selection.activeObject`, indicates which object to use for retrieving packed assets from `AssetBundle` using the `AssetBundle.mainAsset` property. You can set this value to `null` if you are not using it. The second variable is array `Object[]` and specifies which assets you need to pack. The third variable is very simple, and it is just a location where you want to save your `AssetBundle` file. By specifically adjusting `BuildAssetBundleOptions` flags, you can dictate to include all dependencies automatically or only include complete assets. Furthermore, with these options, you can specify that you don't want your assets to be compressed in the `AssetBundle` file by setting up the `UncompressedAssetBundle` flag. If you want, you can check the CRC checksum of your asset bundles while downloading your exported files via the `WWW.LoadFromCacheOrDownload` call. To create asset bundles, you can use three different functions:

- `BuildPipeline.BuildAssetBundle`: This will build any asset type.
- `BuildPipeline.BuildStreamedSceneAssetBundle`: This will include only scenes.
- `BuildPipeline.BuildAssetBundleExplicitAssetNames`: This is similar to `BuildPipeline.BuildAssetBundle`, but with a little difference. In this function, you can indicate your string identifier for all included objects.

 You can create the asset bundles for the Web-Player platform and use them in a standalone platform and vice versa. Alternatively, for mobile platforms, you can use only their built files. For example, you can create the asset bundles for Android devices and use them only for Android platforms, but you cannot use them for iOS platforms and vice versa. With built files for the iOS platform, you can use them only within iOS boundaries.

In order to use your class to create the asset bundle, you need to do two simple steps:

1. You should select one single or many assets in your project, which will be packed into your `AssetBundle`.

2. From the Unity menu, you should navigate to **PacktPub | AssetBundles | Build Asset Bundle**.

The finished script to create `AssetBundle` should be similar to the code shown here:

```
using UnityEngine;
using UnityEditor;

public class BuilderAssetBundle {
  [MenuItem("PacktPub/AssetBundles/Build Asset Bundle")]
  static void Build() {
    BuildPipeline.BuildAssetBundle(
      Selection.activeObject,
      Selection.GetFiltered(
        typeof(Object),
        SelectionMode.DeepAssets
      ),
      "Assets/Your/Path/To/YourAssetBundle.unity3d",
      BuildAssetBundleOptions.CollectDependencies |
      BuildAssetBundleOptions.CompleteAssets
    );
  }
}
```

We've realized the asset bundles creation, now let's explore how to use them from your scripts. To use your asset bundles, you should follow the next two simple steps:

1. Download your `AssetBundle` files from any local storage like hard drive or remote storage like any web-server, Unity provides the www helper class for such issues.

2. Load or, in other words, unpack your assets from your `AssetBundle` files in order to use them further in your game.

In the following code, we intend to understand the case of using `AssetBundle`. First of all, we need to create a simple C# script and name it, for example, `ImporterAssetBundle`. Open the script in your code editor and change it as shown here:

```
using UnityEngine;
using System.Collections;
```

```
public class ImporterAssetBundle : MonoBehaviour {
  void Start() {
    StartCoroutine(Import());
  }

  public IEnumerator Import() {
    using (WWW wwwData = WWW.LoadFromCacheOrDownload(
        "http://your-domain.com/your/path/url/new2.unity3d",
        23 // your asset bundle version, as an example only
    )) {
      yield return wwwData;
      GameObject obj = www.assetBundle.mainAsset as GameObject;
      Instantiate(obj);
      www.assetBundle.Unload(false);
    }
  }
}
```

Attach this script as a component to your GameObject. The URL in this script is just for example, so you should set your own URL for your AssetBundle file. Based on this version, Unity's caching system will decide whether to download your file or not. If this file was already cached with the same version number, then Unity will speed up your application. In this script, we called Import co-routine in the Start event, but you can call this function anywhere you need and as often as you want. In Import co-routine, we first used WWW.LoadFromCacheOrDownload with the sole purpose of loading the required AssetBundle file by given the URL path and the version number. During the downloading process, Unity will not execute the next instructions from the yield return www command. Only after the download is complete will Unity run the next commands in our co-routine example.

- Using the www instance to retrieve mainAsset as a GameObject instance from the downloaded AssetBundle

- Creating a new instance in real time in the current scene of the retrieved GameObject instance

- Unloading all the memory used for this **AssetBundle** (for increasing the performance).

As making projects is an iterative procedure, you will probably alter your assets not once but multiple times, which may oblige remaking the asset bundles after every change to have the capacity to test them. Despite the fact that it is conceivable to load the asset bundles in the Unity editor, it is not the best solution. Rather, while testing in the Unity editor, you ought to utilize the aide function `Resources.LoadAssetAtPath` to abstain from needing to utilize and revamp the asset bundles. The function gives you a chance to load your asset as though it were being stacked from an asset bundle, yet we will skirt the building procedure, and your assets are constantly up and coming. Let's create a new C# script that will improve our last example with some exception handling and with `Resources.LoadAssetAtPath` in event of importing within the Unity editor. Create a new script and name it, for example, `MyAssetBundleImporter`. The next step is to declare at the beginning of the file the required namespaces that we will use:

```
using UnityEngine;
using System.Collections;
After these lines we will declare our public class:
public class MyAssetBundleImporter {
  // the code you will see below
}
```

Let's declare the `public` property in this class for objects that will be retrieved from your `AssetBundle`:

```
public Object assetBundleObject;
```

Also, in this class, let's declare our `public` and specific `AssetBundle` structure with its public properties for further usage:

```
public struct AssetBundleStruct {
  public string assetSourceName;
  public string assetSourcePath;
  public string assetBundleUrl;
  public int assetBundleVersion;
}
```

Next, we will declare the core of this script, the function which Unity will execute as a co-routine:

```
public IEnumerator Import<T>(AssetBundleStruct abs) where T : Object
{}
```

The first step in the co-routine is to initialize our `assetBundleObject` to `null`:

```
assetBundleObject = null;
```

After that line of code, let's declare the main condition in order to decide how to import your asset bundles:

```
#if UNITY_EDITOR
  // 1st part, the code you will see below
#else
  // 2nd part, the code you will see below
#endif
```

Let's create a code for the first part of the if/else/endif pre-processor statement:

```
assetBundleObject = Resources.LoadAssetAtPath(
  abs.assetSourcePath, typeof(T)
);

if (null == assetBundleObject){
  Debug.LogError("AssetBundle ERROR Path: " + abs.assetSourcePath);
  Debug.LogError("Asset Bundle could not be found !!!");
}
yield break;
```

Now, we will create the code for the second part of the if/else/endif preprocessor statement:

```
WWW www;
if (Caching.enabled) {
  while (false == Caching.ready) {
    yield return null;
  }
  www = WWW.LoadFromCacheOrDownload(
    abs.assetBundleUrl, abs.assetBundleVersion
  );
} else {
  www = new WWW(abs.assetBundleUrl);
}

yield return www;

if (null != www.error) {
  Debug.LogError(www.error);
  www.Dispose();
  yield break;
}
```

```
    AssetBundle ab = www.assetBundle;
    www.Dispose();
"   www = null;
    if (string.Empty == abs.assetSourceName || null == abs.
assetSourceName) {
      assetBundleObject = ab.mainAsset;
    } else {
      assetBundleObject = ab.Load(abs.assetSourceName, typeof(T));
    }

    ab.Unload(false);
```

Let's create a simple script inherited from MonoBehaviour, which will show how you can use your example class MyAssetBundleImporter:

```
using UnityEngine;
using System.Collections;

public class ExampleMyImporterUsage : MonoBehaviour {
  public MyAssetBundleImporter.AssetBundleStruct abs;

  private string _tmpStr;
  private Object _tmpObj;
  void Start() {
    abs = new MyAssetBundleImporter.AssetBundleStruct();
    abs.assetBundleUrl = "http://yourapp.com/your/bundle.unity3d";
    abs.assetBundleVersion = 0;
    abs.assetSourceName = "YourPrefabName";

    StartCoroutine(Import());
  }

  IEnumerator Import() {
    MyAssetBundleImporter mabi = new MyAssetBundleImporter();
    yield return StartCoroutine(mabi.Import<GameObject>(abs));
    if (null != mabi.assetBundleObject) {
      _tmpObj = Instantiate(mabi.assetBundleObject);
    }
  }

  void OnGUI() {
    if (null != _tmpObj) {
      _tmpStr = _tmpObj.name + " was successfully created.";
```

```
      GUILayout.Label(_tmpStr);
    } else {
      GUILayout.Label("ERROR: Cannot import your AssetBundle.");
    }
  }
}
```

Let's describe what happened in the last script. First, we declared one public `AssetBundleStruct` variable and two private variables. After that, we created the `Start` method, where we initialized our `AssetBundleStruct` variable with the right values. Next, we called the co-routine `Import` in our `Start` function. In the `Import` co-routine, we created one single instance of our class `MyAssetBundleImporter` to call its `Import` co-routine with our initialized structure abs. If we imported the object and it is not equal to `null`, then we instantiate that GameObject in our scene. Also, we show the simple GUI label that indicates whether we successfully imported our `AssetBundle` or not. For using this (only if your `AssetBundle` was already created and uploaded) script, you should just attach it as a component to your GameObject in your scene and set up correct values. If those two simple steps were done, then you can play and test your game in the Unity editor or within your build.

If you need to get an array with all contained objects from your `AssetBundle`, you should use the function known as `AssetBundle.LoadAll`. In the event where you need to get a list of the string identifiers, you should keep a specific `TextAsset` as a map to save your assets' names there.

In the following steps, we're going to show a simple example about adjusting different texture compressions before building your asset bundle. All we need to remember is two simple steps:

1. In order to force reimporting your assets before building the asset bundle, you should use the `AssetDatabase.ImportAsset` function.

2. After you should use `AssetPostprocessor.OnPreprocessTexture` to adjust correct values for your texture compression.

Now, let's write a simple code example that you can use in your projects like any other examples from this book:

```
using UnityEngine;
using UnityEditor;
```

As always, we declare the required namespaces (remember this script will use `UnityEditor` and should be located in `Editor` folder, we spoke earlier about this Unity requirement). The next step is to define a simple class using the following code:

```
public class TextureFormatAssetBundles {
  // the code you will see below
}
```

After the simple class declaration, we need to set up one `public` and `static` variable, at the same time, and three different but elementary functions (you can use any desired texture format; in this example, we use `DXT1`, `DXT5`, and `ETC_RGB4`):

```
public static TextureImporterFormat tif;

[MenuItem("PacktPub/AssetBundles/Create Asset Bundle DXT1")]
static void SetTextureFormatDXT1() {
  tif = TextureImporterFormat.DXT1;
  CreateAssetBundle();
}

[MenuItem("PacktPub/AssetBundles/Create Asset Bundle DXT5")]
static void SetTextureFormatDXT5() {
  tif = TextureImporterFormat.DXT5;
  CreateAssetBundle();
}

[MenuItem("PacktPub/AssetBundles/Create Asset Bundle ETC_RGB4")]
static void SetTextureFormatETC_RGB4() {
  tif = TextureImporterFormat.ETC_RGB4;
  CreateAssetBundle();
}
```

Now, we can write our main function `CreateAssetBundle` that does all the dirty work:

```
static void CreateAssetBundle() {
  // the code you will see below
}
```

In our first step, we call the `EditorUtility.SaveFilePanel` method in order to show the Unity dialog and to get a selected path string from it. Also, we need to return from this function an empty location variable:

```
string selectedPath = EditorUtility.SaveFilePanel(
  "Save", // TITLE
  string.Empty, // DIRECTORY PATH
```

```
  "Your AssetBundle Name", // DEFAULT FILE NAME
  "unity3d" // FILE EXTENSION
);

if (selectedPath.Length == 0) return;
```

The next step is to declare an array with our selected objects:

```
Object[] selectedAssets = Selection.GetFiltered(
  typeof(Object), SelectionMode.DeepAssets
);
```

Further, we have to process each texture from our array in a loop in order to get path for that asset source with the help of the `AssetDatabase.GetAssetPath` method. For texture, we must use the `AssetDatabase.ImportAsset` function to force our texture preprocessing:

```
for (int i=0; i < selectedAssets.Length; i++) {
  Object obj = selectedAssets[i];
  if ((obj is Texture) == false) continue;
  string texturePath = AssetDatabase.GetAssetPath(
    (UnityEngine.Object) obj
  );
  AssetDatabase.ImportAsset(texturePath);
}
```

After this, we have to build our `AssetBundle`:

```
BuildPipeline.BuildAssetBundle(
  Selection.activeObject,
  selectedAssets,
  selectedPath,
  BuildAssetBundleOptions.CollectDependencies |
  BuildAssetBundleOptions.CompleteAssets
);
```

At the last stage, we can initialize the `Selection.objects` list variable with our selected assets array in order to see all of them:

```
Selection.objects = selectedAssets;
```

Now, we should create a very simple class (this script should be placed into the `Editor` folder, we mentioned this requirement earlier) inherited from the `AssetPostprocessor` class as shown here:

```
using UnityEngine;
using UnityEditor;
```

```
public class TextureAssetsPreprocessor : AssetPostprocessor {
  void OnPreprocessTexture() {
    TextureImporter ti = assetImporter as TextureImporter;
    ti.textureFormat = TextureFormatAssetBundles.tif;
  }
}
```

Downloading new code and data in real time for Android devices

In the event of retrieving assets from your bundles, you can use three separate functions:

- AssetBundle.Load: This will load one object only by the given name; also it will block the main thread.

- AssetBundle.LoadAsync: This will load one object only by a given name; it will not block the main thread. Use this method for huge assets.

- AssetBundle.LoadAll: This will load every object from your AssetBundle.

Use the AssetBundle.Unload method in the event of unloading assets. Let's look at a simple usage example of the asynchronous method as shown in the following code without any exception handling and any checks (just as skeleton):

```
using UnityEngine;
using System.Collections;

public class GetAssetBundleAsync : MonoBehaviour {
  public string assetBundleUrl = "http://yourweb.com/yourBundle.
unity3d";
  public int assetBundleVersion = 1;

  IEnumerator Start() {
  WWW www = WWW.LoadFromCacheOrDownload(
    assetBundleUrl, assetBundleVersion
  );

  yield return www;

  AssetBundle ab = www.assetBundle;

  AssetBundleRequest abr = ab.LoadAsync(
    "YourObjName", typeof(GameObject)
```

```
    );

    yield return abr;

    GameObject go = abr.asset as GameObject;

    ab.Unload(false);
    www.Dispose();
  }
}
```

Managing loaded asset bundles

Asset bundles cannot be loaded if a previous bundle has not been unloaded previously:

```
AssetBundle ab = www.assetBundle;
```

 Try to always keep references for your imported assets, to avoid importing same assets multiple times.

Unity will throw an exception and your asset bundle (in our example, `ab` variable) variable will be `null`.

 Try to unload your `AssetBundle` as early as possible.

You can use the next simple script (as shown in the following code) for your loaded bundles. All this code should be pretty clear to you, so let's look at this C# script:

```
using UnityEngine;
using System;
using System.Collections;
using System.Collections.Generic;

static public class YourAssetBundleDispatcher {
  static Dictionary<string, YourBundleReference> dictionaryBundles;

  static YourAssetBundleDispatcher() {
        dictionaryBundles = new Dictionary<string,
YourBundleReference>();
  }
```

```
private class YourBundleReference {
    public AssetBundle ab = null;

    public int assetBundleVersion;
    public string assetBundleUrl;

    public YourBundleReference(string url, int version) {
        assetBundleUrl = url;
        assetBundleVersion = version;
    }
};

public static AssetBundle
    RetrieveAssetBundle(string abUrl, int abVersion) {
    string bundleKey = abUrl + abVersion.ToString();

YourBundleReference ybr;

if (dictionaryBundles.TryGetValue(bundleKey, out ybr))
        return ybr.ab;
    else
        return null;
 }

  public static IEnumerator ImportAssetBundle(string abUrl, int
abVersion){
      string bundleKey = abUrl + abVersion.ToString();

  if (dictionaryBundles.ContainsKey(bundleKey)) {
          yield return null;
      } else {
          using(WWW www = WWW.LoadFromCacheOrDownload(abUrl,
abVersion)){
              yield return www;

              if (www.error != null)
                  throw new Exception("WWW ERROR:" + www.
error);
      YourBundleReference ybr = new YourBundleReference(
        abUrl, abVersion
      );
              ybr.ab = www.assetBundle;
              dictionaryBundles.Add(bundleKey, ybr);
          }
```

```
        }
    }

    public static void Dispose(string abUrl, int abVersion, bool flag) {
        string bundleKey = abUrl + abVersion.ToString();

        YourBundleReference ybr;

            if (dictionaryBundles.TryGetValue(bundleKey, out ybr)){
                ybr.ab.Unload(flag);
                ybr.ab = null;
                dictionaryBundles.Remove(bundleKey);
            }
        }
    }
```

You can use the dispatcher class as shown here:

```
using UnityEngine;
using System.Collections;

class DispatcherUsage : MonoBehaviour {
  public string assetBundleUrl;
    public int assetBundleVersion;

    AssetBundle ab;

    void Start() {
    Debug.Log("Importing your Asset Bundle");
            ab = YourAssetBundleDispatcher.RetrieveAssetBundle(
      assetBundleUrl, assetBundleVersion
    );
            if(null != ab) StartCoroutine(ImportAssetBundle());
    }

    IEnumerator ImportAssetBundle() {
        yield return StartCoroutine(
      YourAssetBundleDispatcher.ImportAssetBundle(
        assetBundleUrl, assetBundleVersion
      )
    );

        ab = YourAssetBundleDispatcher.RetrieveAssetBundle(
      assetBundleUrl, assetBundleVersion
    );
```

```
        }

    void OnDisable() {
            YourAssetBundleDispatcher.Dispose(
          assetBundleUrl, assetBundleVersion, false
        );
    }
}
```

 It is possible to clone your previous instantiated objects to avoid unnecessary importing bundles (by calling Unity's function known as GameObject.Instantiate).

Asset bundles and binary data

Unity treats binary files with a .bytes extension as a TextAsset, which can be included in your AssetBundle. Here's an example of the C# script shown here:

```csharp
using UnityEngine;
using System.Collections;

public class BinaryDataExample : MonoBehaviour {
  string assetBundleUrl =
    "http://yourweb.com/path/to/yourAssetBundle_1.unity3d";

  IEnumerator Start() {
    WWW www = WWW.LoadFromCacheOrDownload(assetBundleUrl, 1);
    yield return www;

    AssetBundle ab = www.assetBundle;

    TextAsset textAsset = ab.Load(
      "YourBinaryFileName", typeof(TextAsset)
    ) as TextAsset;

    byte[] yourBinaryData = textAsset.bytes;
  }
}
```

Asset bundles and scripts

You can build your asset bundles with scripts as `TextAsset` files, which can be executable only in the event of precompiling them into an assembly. This example is shown in the following code:

```
using UnityEngine;
using System.Collections;

public class AssetBundleScript : MonoBehaviour {
  string assetBundleUrl =
    "http://yourweb.com/your/asset/bundle_test.unity3d";

  IEnumerator Start () {
    WWW www = WWW.LoadFromCacheOrDownload (assetBundleUrl, 1);
      yield return www;

      AssetBundle ab = www.assetBundle;

      TextAsset textAsset = ab.Load(
    "yourBinaryAssetName", typeof(TextAsset)
  ) as TextAsset;

      var assmbl = System.Reflection.Assembly.Load(textAsset.bytes);

  GameObject gameObj = new GameObject();
      gameObj.AddComponent(
    assmbl.GetType(
      "Your_ClassName_Inherited_From_MonoBehaviour"
    )
  );
  }
}
```

Asset bundle dependencies

Many of your assets will depend on other assets, such as materials, textures, shaders, and so on. You can build your bundle with all those assets, but this approach can reduce the size of the `AssetBundle` file. Furthermore, this approach will not be effective if all those dependencies are used for your other bundles. Too much memory will be wasted. Instead, you can create a separate asset bundle with all those shared dependencies, which will be used by other bundles. In the event of using these dependencies, firstly, you should call the function known as `BuildPipeline.PushAssetDependencies`, and then your shared bundle can be built for other bundles. Therefore, before each new level, you should always call this function to tell Unity to put your bundle in its stack for further usage by other bundles. At the end of your bundle creation, you should always empty this stack of bundles by a command known as `BuildPipeline.PopAssetDependencies`. In your application, you should always import all your shared bundles and only after that you can import your other bundles with those dependencies. Let's see how to save separate `AssetBundle` with shared shaders (as shown here) in this action:

```
using UnityEngine;

public class YourAssetBundleShaders : MonoBehaviour {
        public Shader[] assetBundleShaders;
}
```

Create an empty GameObject and attach this script to it and save it (after populating array of shaders) as a prefab anywhere you want within your project files. The following step is to create the C# script to generate asset bundles as shown here, where `YourAssetBundle_2` requires `YourAssetBundle_1` and `YourAssetBundle_3` depends on the first and the second bundles. This is just an example, you should change it depending on your needs:

```
using UnityEngine;
using UnityEditor;

public class AssetBundleGenerator {
   [MenuItem("PacktPub/AssetBundles/Generate all accessible Asset
Bundles")]
        static void Generate() {
          BuildAssetBundleOptions options =
                  BuildAssetBundleOptions.CollectDependencies |
                  BuildAssetBundleOptions.CompleteAssets |
                  BuildAssetBundleOptions.DeterministicAssetBundle;

     BuildPipeline.PushAssetDependencies();
```

```
        BuildPipeline.BuildAssetBundle(
    AssetDatabase.LoadMainAssetAtPath(
      "Assets/YourAssetName_1.prefab"
    ),
    null,
    "Your/Path/To/YourAssetBundle_1.unity3d",
    options
  );

  BuildPipeline.PushAssetDependencies();
  BuildPipeline.BuildAssetBundle(
    AssetDatabase.LoadMainAssetAtPath(
      "Assets/YourAssetName_2.prefab"
    ),
    null,
    "YourPath/To/YourAssetBundle_2.unity3d",
    options
  );

  BuildPipeline.BuildAssetBundle(
    AssetDatabase.LoadMainAssetAtPath(
      "Assets/YourAssetName_3.prefab"
    ),
    null,
    "YourPath/To/YourAssetBundle_3.unity3d",
    options
  );

  BuildPipeline.PopAssetDependencies();
  BuildPipeline.PopAssetDependencies();
}

[MenuItem("PacktPub/AssetBundles/Rebuild Asset Bundle")]
    static void Rebuild() {
  BuildAssetBundleOptions options =
    BuildAssetBundleOptions.CollectDependencies |
    BuildAssetBundleOptions.CompleteAssets |
    BuildAssetBundleOptions.DeterministicAssetBundle;

  BuildPipeline.PushAssetDependencies();
  BuildPipeline.BuildAssetBundle(
    AssetDatabase.LoadMainAssetAtPath(
      "Assets/YourAssetBundleName_1.prefab"
    ),
```

```
    null,
    "YourPath/To/YourAssetBundle_1.unity3d",
    options
  );

  BuildPipeline.PopAssetDependencies();
}
}
```

Safeness techniques in practice

Next, the C# script example will cover how to protect the content of your asset bundles (as shown here):

```
using UnityEngine;
using System.Collections;

public class AssetBundleSecurityFirst : MonoBehaviour {
  string assetBundleUrl =
    "http://yourweb.com/path/to/yourAssetBundle.unity3d";

  IEnumerator Start() {
    WWW www = WWW.LoadFromCacheOrDownload(assetBundleUrl, 1);
    yield return www;

    TextAsset textAsset = www.assetBundle.Load(
      "YourEncryptedAssetName", typeof(TextAsset)
    ) as TextAsset;

    /*byte[] yourDecryptedBytes = AnyDecryptionFunction(
      textAsset.bytes // your encrypted bytes
    );*/
  }
}
```

Another secure way is to encrypt the whole asset bundle instead of just the TextAsset data as shown in the preceding code. Alternatively, in this approach, you cannot use the WWW.LoadFromCacheOrDownload method. You always need to import your bundles from WWW streaming as shown in the following code:

```
using UnityEngine;
using System.Collections;

public class AssetBundleSecuritySecond : MonoBehaviour {
```

```
    string assetBundleUrl =
      "http://yourweb.com/path/to/yourAssetBundle.unity3d";

  IEnumerator Start () {
    WWW www = new WWW(assetBundleUrl);
    yield return www;

    /*byte[] yourEncryptedBytes = www.bytes;
    byte[] yourDecryptedBytes =
      AnyDecryptionFunction(yourEncryptedBytes);

    AssetBundleCreateRequest assetBundleCreateRequest =
      AssetBundle.CreateFromMemory(yourDecryptedBytes);

    yield return assetBundleCreateRequest;

    AssetBundle ab = assetBundleCreateRequest.assetBundle;*/

    // Here you can use your AssetBundle.
    // The AssetBundle was not cached.
  }
}
```

The last and the best protection approach is to keep your encrypted `AssetBundle` as `TextAsset` inside another (not encrypted) `AssetBundle`. Thus, we can use Unity's caching system for our asset bundles as shown here:

```
using UnityEngine;
using System.Collections;

public class AssetBundleSecurityThird : MonoBehaviour {
  string assetBundleUrl =
    "http://yourweb.com/path/to/yourAssetBundle.unity3d";

  IEnumerator Start() {
    WWW www = WWW.LoadFromCacheOrDownload(assetBundleUrl, 1);
    yield return www;

    TextAsset textAsset = www.assetBundle.Load(
      "YourEncryptedAsset", typeof(TextAsset)
    ) as TextAsset;

    /*byte[] yourEncryptedBytes = textAsset.bytes;
    byte[] yourDecryptedBytes =
```

```
        AnyDecryptionFunction(yourEncryptedBytes);

    AssetBundleCreateRequest assetBundleCreateRequest =
        AssetBundle.CreateFromMemory(yourDecryptedBytes);
    yield return assetBundleCreateRequest;

    AssetBundle ab = assetBundleCreateRequest.assetBundle;*/
    // Here you can use your AssetBundle. The AssetBundle was cached.
    }
}
```

Summary

In this chapter, we discovered asset bundles. There were a lot of code examples. We learned how to build and import your asset bundles and how to encrypt data for the asset bundle by different methods. Furthermore, in this chapter, we explored how to create and use your AssetBundle dependencies with other bundles. Also, we studied how to use binary data with asset bundles and executable scripts.

In the next chapter, we will introduce different optimization techniques. You will learn in practice how to use occlusion culling and level of details optimization techniques. You will see how to optimize native C# and Unity scripts. Finally, you will see how to transform native C# and JavaScript codes into Unity scripts.

6
Optimization and Transformation Techniques

This chapter will introduce you to the usage of **occlusion culling** (OC) and level of detail in optimization techniques. Also, you will learn to optimize Unity C# and Unity JS code. Finally, you will see how to transform Unity C# code to Unity JavaScript code and vice versa.

The topics that will be covered in the chapter are as follows:

- Occlusion culling and level of detail in optimization techniques
- Unity C# and Unity JS optimization tips and tricks
- Transforming Unity C# code to Unity JavaScript code and vice versa

Occlusion culling and level of detail in optimization techniques

Let's look more closely and carefully at the basic principles of occlusion culling in Unity (only Pro license) and how to use them in your projects to achieve excellent performance.

You can open the occlusion culling editor from the Unity menu as shown in the following screenshot:

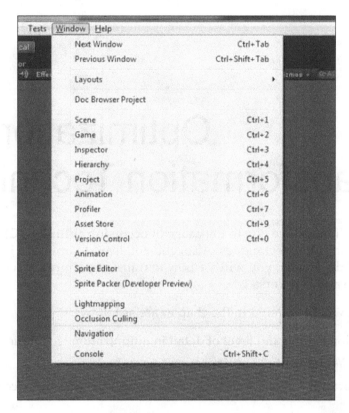

The main aim of the occlusion culling mechanism is sifting and filtering off objects that are not visible in the camera area in order to improve optimization. This primarily means that the objects will not be using resources, only when necessary, with the result that of helping you create a game or app that will work much faster.

Frustum culling is different from occlusion culling because it disables the renderers that are outside the view of the camera, but does not disable the renderers that overlap other renderers; for example, if a wall hides an object, it will be invisible for the camera. Using occlusion culling, you can automatically take the advantage of frustum culling. With the usage of visual occlusion culling technique, we can see in two examples as shown in the following screenshots:

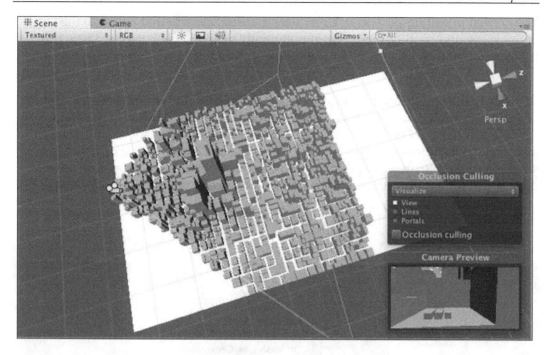

In the screenshot shown here, you can see occlusion culling in action:

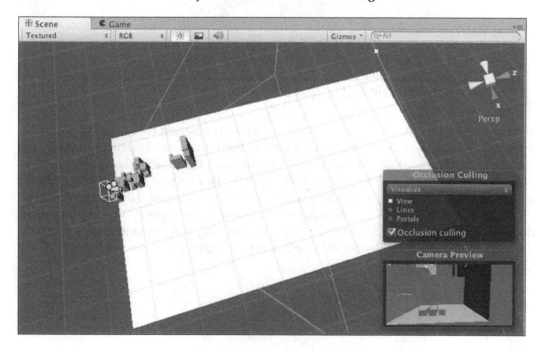

The occlusion culling process in Unity uses a virtual camera that will scan the entire scene and create a hierarchy of potentially visible sets of objects. Then, this information will be used by other cameras in your game or in your application in real time, in order to reduce the number of draw calls and to improve your performance.

In order to use occlusion culling, you need to set the **Occluder Static** tag for each object in the scene to be processed by this optimization mechanism. Also, you can use another object's tag, which is called **Occludee Static** as shown in the next screenshot. Occludees may be obscured by other objects and will be disabled in a similar situation to improve performance, but these objects cannot overlap other objects. Therefore, they will increase the performance of your entire project.

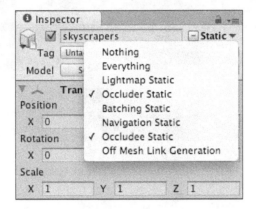

 It is also very important to create areas for occlusion culling, only where the camera will render objects.

We just opened the basic and key aspects of the optimization by occlusion culling approach. We cannot describe all the details of the settings and features in this chapter. The following sections describe the various ideas, methods, approaches, and ways to optimize and improve performance. The purpose of the next section is to direct you on the right path to improve performance. You will use the desired optimization techniques from this chapter, and if it is necessary, you can find more detailed information about methods, implementations, and customizations from the web. Let's now consider another optimization technique known as **Level Of Detail (LOD)**.

Optimizing by LOD

The LOD optimization technique is a method of reducing the complexity of frame rendering by reducing the total number of polygons, textures, and other resources in the scene, the general decline in its complexity. A simple example is that the main character model consists of 10,000 polygons. In cases where the treated stage is located close to the camera, it is important to use all the polygons. However, at a great distance from the camera, in the final image, it will take only a few pixels; there is no sense in handling all 10,000 polygons. Perhaps, in this case, it would be enough for hundreds of polygons, or even a couple of pieces and textures, specially prepared for about the same display model. Accordingly, at intermediate distances, it makes sense to use a model consisting of a number of triangles greater than the simplest model and smaller than the most complex.

The LOD method is commonly used for modeling and rendering three-dimensional scenes using multiple difficulty levels (geometric or some other) for the objects in proportion to their distance from the camera. Changing complexity, particularly in the number of triangles in the model may be performed automatically, is based on a three-dimensional model of highest complexity, but can be on the basis of several predefined patterns with different levels of detail. Using a model with less detail for different distances, you will reduce rendering design complexity, almost without compromising the overall image detail.

The method is particularly effective when the number of objects in the scene is large, and they are located at different distances from the camera. For example, consider a sports game, such as football game or a hockey simulator. Low-poly character models are used when they are away from the camera, but when it approaches, the models are replaced by a large number of polygons. This example is very simple, and it shows that the essence of the method is based on two levels of detail of the model, but no one bothers to create multiple levels of detail. In order to effect change, the LOD level was not too obvious, so the object detail gradually grows.

Consider the following factors that have an effect on the level of detail technique: the total number of objects on the screen (when one or two characters in the frame, use complex models, and when 10–20, they are switched to a simpler model) or the FPS (predetermined limited values of the FPS, which varies with the level of detail, such as FPS below 30 reduces the complexity of the models on the screen, while 60 FPS raises the complexity). Other possible factors that affect the level of detail are as follows: the speed of movement of the object (in case of a rocket in motion you see it moving fast, but a snail moves slowly), the importance of a character from the game's point of view (for example, in football, the player models you see the closest and most often uses more complex geometry and texture). It all depends on the desires and capabilities of a particular developer. The main thing is to not overdo it; frequent and noticeable changes in the level of detail will be annoying.

We want to remind you that the level of detail does not necessarily refer only to the geometry. The method can also be used to save other resources: texturing (although GPUs use mipmapping, sometimes it makes sense to change the texture on the fly on the other with some detail), lighting technician (close objects covered by a complex algorithm, and distant objects covered by a prime), and texturing techniques.

Unity C# and Unity JS optimization tips and tricks

To begin, we will consider some aspects of optimization concerning the JavaScript programming language. Try to avoid the use of dynamic typing in JavaScript. The best solution for your performance is undoubtedly static typing. The use of dynamic typing of variables will be consumed while executing a code to find the appropriate data type for a particular variable, which in principle could and should be avoided by specifying the data types for all your variables. Bad and good examples are shown here:

```
// Dynamic Typing, BAD FOR YOUR PERFORMANCE
var yourVariableName = 23;


// Static Typing, GOOD FOR YOUR PERFORMANCE
var myGo : GameObject = null;
```

The following example shows what you should not do if you want to improve your performance. This example uses dynamic typing for our variable, yourVariableName, which in turn affects the performance of the whole system in the negative sense. Before calling any function of this object, there will be time spent searching for the right object type and checking if the called function is accessible. The bad example is shown here:

```
function Start() {
  var yourVariableName = GetComponent(YourScriptName);
  yourVariableName.YourFunctionName();
}
```

Instead of wasting CPU time on unnecessary expenses, you should always use static typing for all your variables in order to improve performance:

```
function Start() {
  var yourVariableName : YourScriptName =
GetComponent(YourScriptName);
  yourVariableName.YourFunctionName();
}
```

Use Static Typing instead of Dynamic Typing wherever possible

You can use #pragma strict preprocessor directive in order to help you not to forget about using static typing instead of dynamic typing everywhere in your JavaScript script. You should write this directive at the top of your script before any code. In the event of utilizing #pragma strict and dynamic typing in your script, the compiler will throw errors. Therefore, this preprocessor directive forces you to use static typing only.

We also want to mention the other techniques to optimize your code. One of them is the technique of caching components or variables. During optimization, you first need to turn your attention to the functions that are very often performed in your code, especially callbacks such as Update and FixedUpdate and similar functions that are called in every frame, or almost each frame, or in other words, many times per second. Therefore, reference to any component or variable in such risky functions depends on the situation. Of course, there are situations for the overall system performance, where such things are not so bad for your performance in general, and there are situations where performance drops almost to zero because of the many unnecessary expenses. In such functions, it is best not to call the Unity method GetComponent each time, which will very often look for a component or other similar functions of the Unity library to find objects and so on. Instead, you can call the function you want, when it is necessary to obtain a desired component or desired object(s) and store them in local variables or arrays, as you like. The following examples demonstrate this:

```
// BAD for your performance
void Update() {
   transform.position = new Vector3(0.0f, 1.0f, -1.0f);
}


// Second example:
// GOOD for your performance
private Transform _t;
void Start() {
  _t = transform;
}

void Update() {
  _t.position = new Vector3(0.0f, 1.0f, -1.0f);
}
```

The code shown in the second example is much faster then the respective code in the first example, because Unity will not look up the transform component each `Update` cycle or in other words each frame.

 You should call a function only if necessary, not more and not less, just exactly when needed.

The best optimization for your code and generally for the whole system is when the code is as small as possible, or rather when nothing unnecessary is executed at all. Unnecessary calculations lead to unnecessary overhead; for mobile devices, in general, this question is one of the most acute. A good example of a small optimization is shown in the following example, but this example is not the best solution for your performance. After each frame, checking the distance between the two points will take away your precious time:

```
private Transform yourTransform:
void Update() {
  if (Vector3.Distance(yourTransform.position, transform.position) >
200) {
    return;
  }
  // your next code may be here ...
}
```

To not waste time on unnecessary mistakes, you should use the `OnBecameInvisible` and `OnBecameVisible` callbacks. With these callbacks, Unity calls in an event in which none of the cameras can see (for `OnBecameInvisible`) or at least one camera sees (for `OnBecameVisible`) your renderer. Certainly, these callbacks are only good in certain situations and not in others. For example, if your object does not contain a renderer component, then you will need to invent a way to enable or disable the execution of your code accordingly. A simple example of these two callbacks is shown here:

```
void OnBecameVisible() {
  enabled = true;
}

void OnBecameInvisible() {
  enabled = false;
}
```

To achieve the performance that you need, you will need to take care of many details in your code and many other details that are discussed in this book. Code optimization in most cases impedes readability and therefore impedes understanding of the code. Keep this in mind, or at least do not forget about it. Like everything else in life, we need to find a middle ground or, in other words, the golden balance between quality and performance.

Let's see how static functions behave and how much time we need for them, since the use of these functions significantly reduces the time of the function call as compared to call a non-static function. If we examine the question of what happens with static functions when you compile your code, it is, however, no secret that all the code is translated into machine code or assembler as it is called, which is the lowest programming level. If we consider the very assembly instructions to call static functions, we see that it requires fewer machine instructions and consequently less CPU time than calling a nonstatic function.

In the call, each function with parameters passed by value requires memory duplication. As discussed earlier in text, this may impair your performance. Therefore, it is better to call a function with parameters passed by reference rather than by value. It is easy to bypass this problem. The best thing for enhancing your performance is to use local variables of a class or object that is used in the function. You can create a set of variables within a function, which in turn will significantly increase the consumption of memory and CPU time of the function call with a lot of variables. The function must be remembered in the stack for further opportunities to work with these variables. Even if these variables are not used in the function, they are still in the stack and they will occupy memory space.

The following discussion focuses on the constants. Constants do not require RAM allocation, since their values are directly sewed in the instruction stream. Using constants instead of creating a large number of local or global variables can significantly accelerate the performance of your software, avoiding the overhead of memory and CPU time.

Static variables (variables of a class) as well as static functions (methods of a class) require less CPU time, since static variables belong to the whole class, rather than to an object of this class. The time spent on searching for supplies is declining, which has obvious advantages in optimization. For variables or functions of any object machine, instructions will be executed to locate the appropriate object to which they belong, which obviously require the overhead of CPU time and memory.

The if and switch statements can be easily changed for each other; for example, to increase the understanding and readability of the code or to optimize all the same code. If you look at the postcompilation machine instructions and directions through any disassembler, you can see the difference between these two expressions. The switch statement, for example, after compiling becomes a go-to mechanism, which in turn makes jumping through machine instructions in its transition table. It needs to find the desired transition in the first place, and then those going to the command assembly. If construction behaves low bit differently, it turns normal branching as in high-level programming language; for example, in our case, the C# language. In some cases, some switch design may be executed faster than the same if/else if/ else design. Performance of these two structures solely depends on their correct application or in other words the correct use. For example, let's consider two simple cases, where in the first case will be quicker and faster switch, and in the second situation the if design is better for performance than a switch solution, as shown in the following code example:

```csharp
using UnityEngine;
using System.Diagnostics;

public class IfSwitchTestFirstCase : MonoBehaviour {
  public const int CYCLES_COUNTER = 100000000;

  bool IfTest(int yourIntegerExample)
  {
    if (yourIntegerExample == 0 || yourIntegerExample == 1) {
      return true;
    }

    if (yourIntegerExample == 2 || yourIntegerExample == 3) {
      return false;
    }

    if (yourIntegerExample == 4 || yourIntegerExample == 5) {
      return true;
    }

    return false;
  }

  bool SwitchTest(int yourIntegerExample)
  {
    switch (yourIntegerExample)
    {
      case 0:
```

```
      case 1:
        return true;

      case 2:
      case 3:
        return false;

      case 4:
      case 5:
        return true;

      default:
        return false;
    }
  }

  void Start() {
    Stopwatch ifTimer = Stopwatch.StartNew();
    for (int i = 0; i < CYCLES_COUNTER; i++)
    {
      IfTest(i);
    }
    ifTimer.Stop();

    Stopwatch switchTimer = Stopwatch.StartNew();
    for (int i = 0; i < CYCLES_COUNTER; i++)
    {
      SwitchTest(i);
    }
    switchTimer.Stop();

    UnityEngine.Debug.Log(
      "IF time = " +
      (
(double)(ifTimer.Elapsed.TotalMilliseconds * 1000 * 1000) / CYCLES_
COUNTER
      ).ToString("0.00 nanoseconds average per cycle")
    );

    UnityEngine.Debug.Log(
      "Switch time = " +
      (
(double)(switchTimer.Elapsed.TotalMilliseconds * 1000 * 1000) /
CYCLES_COUNTER
```

```
    ).ToString("0.00 nanoseconds average per cycle")
  );
}
}
```

On Mac OS X, Intel Core i5 2.7 GHz, after testing the first example in the Unity editor the results were:

```
IF time = 11.54 nanoseconds average per cycle
Switch time = 8.76 nanoseconds average per cycle
```

Based on the preceding results, we can say that the design of switch affects your performance better, but it is not always true. Let's now consider the second case, where the if construction turns out to be the best design solution for your optimization, as shown in code example here:

```
using UnityEngine;
using System.Diagnostics;

public class IfSwitchTestSecondCase : MonoBehaviour {
  public const int CYCLES_COUNTER = 100000000;

  int SwitchTest(int yourIntegerExample)
  {
    switch (yourIntegerExample)
    {
      case 0:
      {
        return 11;
      }

      case 1:
      {
        return 22;
      }

      default:
      {
        return -11;
      }
    }
  }

  int IfTest(int yourIntegerExample)
  {
```

```
  if (0 == yourIntegerExample)
  {
    return 11;
  }

  if (1 == yourIntegerExample)
  {
    return 22;
  }

  return -11;
}

void Start() {
  Stopwatch switchTimer = Stopwatch.StartNew();
  for (int i = 0; i < CYCLES_COUNTER; i++)
  {
    SwitchTest(0);
    SwitchTest(0);
    SwitchTest(0);
    SwitchTest(0);
    SwitchTest(0);
    SwitchTest(0);
    SwitchTest(1);
    SwitchTest(1);
    SwitchTest(1);
    SwitchTest(1);
  }
  switchTimer.Stop();

  Stopwatch ifTimer = Stopwatch.StartNew();
  for (int i = 0; i < CYCLES_COUNTER; i++)
  {
    IfTest(0);
    IfTest(0);
    IfTest(0);
    IfTest(0);
    IfTest(0);
    IfTest(0);
    IfTest(1);
    IfTest(1);
    IfTest(1);
    IfTest(1);
  }
```

```
    ifTimer.Stop();

    UnityEngine.Debug.Log(
       "IF time = " +
       (
(double)(ifTimer.Elapsed.TotalMilliseconds * 1000 * 1000) / CYCLES_
COUNTER
       ).ToString("0.00 nanoseconds average per cycle")
    );

    UnityEngine.Debug.Log(
       "Switch time = " +
       (
(double)(switchTimer.Elapsed.TotalMilliseconds * 1000 * 1000) /
CYCLES_COUNTER
       ).ToString("0.00 nanoseconds average per cycle")
    );
  }
}
```

On a Mac OS X, Intel Core i5 2.7 GHz, after testing the second example in the Unity editor, the results were:

```
IF time = 54.46 nanoseconds average per cycle
Switch time = 64.24 nanoseconds average per cycle
```

Since different situations require different designs, the most important thing is for you to understand the true meaning of what is happening in the construction machine after compiling your code. Then, it will be much easier to make the right choices to improve and enhance your performance. In both examples, discussed earlier, we saw that in different situations the `if` and `switch` designs, which at first sight are absolutely equivalent, may differ in speed and efficiency with respect to performance. We also saw that different situations give a performance advantage with different designs, although they have the same semantics, or in other words, the same algorithm designed in different forms. However, the meaning does not change when dealing with any problems, except for the time of execution, as we have already considered earlier.

The following two-dimensional arrays can be used in the form of one-dimensional arrays, and this will increase your performance. For example, we have a two-dimensional array with **N** rows and **M** of columns: the table size is $N \times M$:

```
// [i, j] from float 2D array (table)
// 0 ≤ i ≤ N - 1
// 0 ≤ j ≤ M - 1
float2Darray[i, j] = 123.321f;
```

In the case of optimization of the two-dimensional array by switching it to the one-dimensional array, we can refer to an element (i, j) of our table with the size N × M as follows:

```
// [i, j] from float 1D array
// 0 ≤ i ≤ N - 1
// 0 ≤ j ≤ M - 1
float1Darray[(i * M) + j] = 123.321f;
```

Here is a complete sample code in Unity C# as shown in the following code example:

```
using UnityEngine;
using System.Diagnostics;

public class Array2Dvs1D : MonoBehaviour {
  public const int N = 1000, M = 1500;

  float[,] float2Darray;
  float[] float1Darray;

  void Start() {
    float2Darray = new float[N, M];
    float1Darray = new float[N * M];

    Stopwatch array2DTimer = Stopwatch.StartNew();
    for (int i = 0; i < N; i++)
    {
      for (int j = 0; j < M; j++)
      {
        // [i, j] from float 2D array
        // 0 ≤ i ≤ N - 1
        // 0 ≤ j ≤ M - 1
        float2Darray[i, j] = 123.321f;
      }
    }
    array2DTimer.Stop();

    Stopwatch array1DTimer = Stopwatch.StartNew();
    for (int i = 0; i < N; i++)
    {
      for (int j = 0; j < M; j++)
      {
        // [i, j] from float 1D array
        // 0 ≤ i ≤ N - 1
        // 0 ≤ j ≤ M - 1
```

```
            float1Darray[(i * M) + j] = 123.321f;
        }
    }
    array1DTimer.Stop();

    UnityEngine.Debug.Log(
        "Array 1D time = " +
        (
(double)(array1DTimer.Elapsed.TotalMilliseconds * 1000 * 1000) / (N *
M)
        ).ToString("0.00 nanoseconds average per cycle")
    );

    UnityEngine.Debug.Log(
        "Array 2D time = " +
        (
(double)(array2DTimer.Elapsed.TotalMilliseconds * 1000 * 1000) / (N *
M)
        ).ToString("0.00 nanoseconds average per cycle")
    );
    }
}
```

On Mac OS X, Intel Core i5 2.7 GHz, after testing this example in the Unity editor the following results were obtained:

```
Array 1D time = 3.24 nanoseconds average per cycle
Array 2D time = 7.87 nanoseconds average per cycle
```

As we can see a difference in the implementation of the same ideas, but in different forms as two-dimensional and one-dimensional arrays. Also consider the next simple example code as shown here, which shows a two-level array in comparison with one-dimensional arrays for speed of execution:

```
using UnityEngine;
using System.Diagnostics;

public class LeveledArray2Dvs1D : MonoBehaviour {
    public const int N = 1000, M = 1500;

    float[][] float2Darray;
    float[] float1Darray;

    void Start() {
        float2Darray = new float[N][];
        float1Darray = new float[N * M];
```

```
for (int i = 0; i < N; i++)
{
  float2Darray[i] = new float[M];
}

Stopwatch array2DTimer = Stopwatch.StartNew();
for (int i = 0; i < N; i++)
{
  for (int j = 0; j < M; j++)
  {
    // [i][j] from float 2D array
    // 0 ≤ i ≤ N - 1
    // 0 ≤ j ≤ M - 1
    float2Darray[i][j] = 123.321f;
  }
}
array2DTimer.Stop();

Stopwatch array1DTimer = Stopwatch.StartNew();
for (int i = 0; i < N; i++)
{
  for (int j = 0; j < M; j++)
  {
    // [i, j] from float 1D array
    // 0 ≤ i ≤ N - 1
    // 0 ≤ j ≤ M - 1
    float1Darray[(i * M) + j] = 123.321f;
  }
}
array1DTimer.Stop();

UnityEngine.Debug.Log(
  "Leveled Array 1D time = " +
  (
(double)(array1DTimer.Elapsed.TotalMilliseconds * 1000 * 1000) / (N *
M)
  ).ToString("0.00 nanoseconds average per cycle")
);

UnityEngine.Debug.Log(
  "Leveled Array 2D time = " +
  (
(double)(array2DTimer.Elapsed.TotalMilliseconds * 1000 * 1000) / (N *
M)
```

```
      ).ToString("0.00 nanoseconds average per cycle")
    );
  }
}
```

On Mac OS X, Intel Core i5 2.7 GHz, after testing this example in the Unity editor the following results were obtained:

```
Leveled Array 1D time = 3.23 nanoseconds average per cycle
Leveled Array 2D time = 3.36 nanoseconds average per cycle
```

For convenience, you can use the two-level array if you are satisfied with its performance. You will need to make the right decision starting from your tasks, not forgetting about the middle ground between the readability of the code and its performance.

As for strings and character arrays, let's see which among them is faster and more efficient. In the following code example, we represent our test performance between the two variables:

```
using UnityEngine;
using System.Diagnostics;

public class StringCharArray : MonoBehaviour {
  public const int LENGTH = 1000;

  string str;
  char[] charArray;

  void Start() {
    charArray = new char[LENGTH];

    Stopwatch charArrayTimer = Stopwatch.StartNew();
    for (int i = 0; i < LENGTH; i++)
    {
      charArray[i] = (i % 10).ToString()[0];
    }
    charArrayTimer.Stop();

    str = string.Empty;
    Stopwatch stringTimer = Stopwatch.StartNew();
    for (int i = 0; i < LENGTH; i++)
    {
      str += (i % 10).ToString();
    }
    stringTimer.Stop();
```

```
    UnityEngine.Debug.Log(
        "String time = " +
        (
(double)(stringTimer.Elapsed.TotalMilliseconds * 1000 * 1000) / LENGTH
        ).ToString("0.00 nanoseconds average per cycle")
    );

    UnityEngine.Debug.Log(
        "Char Array time = " +
        (
(double)(charArrayTimer.Elapsed.TotalMilliseconds * 1000 * 1000) /
LENGTH
        ).ToString("0.00 nanoseconds average per cycle")
    );
  }
}
```

On Mac OS X, Intel Core i5 2.7 GHz, after testing this example in the Unity editor the following results were obtained:

```
String time = 1274.00 nanoseconds average per cycle
Char Array time = 369.00 nanoseconds average per cycle
```

The difference is obvious, but readability of this optimization falls down rapidly. As always everything in life needs a strong balance sheet or, in other words, the golden mean. Here is another example, in which we compare the performance of StringBuilder and character array as shown in the following code example:

```
using UnityEngine;
using System.Text;
using System.Diagnostics;

public class StringBuilderCharArray : MonoBehaviour {
  public const int LENGTH = 1000;

  StringBuilder str;
  char[] charArray;

  void Start() {
    charArray = new char[LENGTH];

    Stopwatch charArrayTimer = Stopwatch.StartNew();
    for (int i = 0; i < LENGTH; i++)
    {
      charArray[i] = (i % 10).ToString()[0];
```

```
    }
    charArrayTimer.Stop();

    str = new StringBuilder();
    Stopwatch stringBuilderTimer = Stopwatch.StartNew();
    for (int i = 0; i < LENGTH; i++)
    {
      str.Append((i % 10).ToString());
    }
    stringBuilderTimer.Stop();

    UnityEngine.Debug.Log(
      "String Builder time = " +
      (
(double)(stringBuilderTimer.Elapsed.TotalMilliseconds * 1000 * 1000) /
LENGTH
      ).ToString("0.00 nanoseconds average per cycle")
    );

    UnityEngine.Debug.Log(
      "Char Array time = " +
      (
(double)(charArrayTimer.Elapsed.TotalMilliseconds * 1000 * 1000) /
LENGTH
      ).ToString("0.00 nanoseconds average per cycle")
    );
  }
}
```

On Mac OS X, Intel Core i5 2.7 GHz, after testing this example in the Unity editor the following results were obtained:

```
String Builder time = 463.00 nanoseconds average per cycle
Char Array time = 370.00 nanoseconds average per cycle
```

StringBuilder is slightly inferior to the performance of a character array. However, do not forget that for the garbage collector, StringBuilder is very well optimized and does not create a memory leak with large volumes of data. You have to solve various problems in the course of software development. If every decision will confidently and firmly take any criticism, then success is not far away. You should definitely prioritize your tasks for all to see where you need to go with compromises.

In the next step, we will examine and study the performance of collections in C#. Collection is very useful in certain situations, but you always have to remember that it is a wrapper for ordinary arrays. When using large data, collections can utilize significant costs of processing time, which in turn negatively affects the whole performance of your code. In the following code example, a list collection execution speed is compared with the conventional one-dimensional array:

```csharp
using UnityEngine;
using System.Collections.Generic;
using System.Diagnostics;

public class ListVsArray : MonoBehaviour {
  public const int LENGTH = 1000000;

  List<int> intList;
  int[] intArray;

  int tmpInt;

  void Start() {
    intList = new List<int>();
    intArray = new int[LENGTH];

    Stopwatch intArrayTimer = Stopwatch.StartNew();
    for (int i = 0; i < LENGTH; i++)
    {
      intArray[i] = i;
      tmpInt = intArray[i]++;
    }
    intArrayTimer.Stop();

    Stopwatch listTimer = Stopwatch.StartNew();
    for (int i = 0; i < LENGTH; i++)
    {
      intList.Add(i);
      tmpInt = intList[intList.Count - 1]++;
    }
    listTimer.Stop();

    UnityEngine.Debug.Log(
      "Integer List time = " +
      (
(double)(listTimer.Elapsed.TotalMilliseconds * 1000 * 1000) / LENGTH
      ).ToString("0.00 nanoseconds average per cycle")
```

```
    );

  UnityEngine.Debug.Log(
    "Integer Array time = " +
    (
(double)(intArrayTimer.Elapsed.TotalMilliseconds * 1000 * 1000) /
LENGTH
    ).ToString("0.00 nanoseconds average per cycle")
  );
  }
}
```

On Mac OS X, Intel Core i5 2.7 GHz, after testing this example in the Unity editor the following results were obtained:

```
Integer List time = 36.68 nanoseconds average per cycle
Integer Array time = 5.54 nanoseconds average per cycle
```

As can be seen from the results discussed earlier in the text, the list collection is significantly inferior in performance compared to simple one-dimensional arrays. Next, as shown in the following code example, the performance of the `ArrayList` class is compared with the same simple one-dimensional array:

```
using UnityEngine;
using System.Collections;
using System.Diagnostics;

public class ArrayListVsArray : MonoBehaviour {
  public const int LENGTH = 1000000;

  ArrayList intArrayList;
  int[] intArray;

  int tmpInt;

  void Start() {
    intArrayList = new ArrayList();
    intArray = new int[LENGTH];

    Stopwatch intArrayTimer = Stopwatch.StartNew();
    for (int i = 0; i < LENGTH; i++)
    {
      intArray[i] = i;
      tmpInt = intArray[i] + 23;
    }
```

```
      intArrayTimer.Stop();

      Stopwatch arrayListTimer = Stopwatch.StartNew();
      for (int i = 0; i < LENGTH; i++)
      {
        intArrayList.Add(i);
        tmpInt = (int)intArrayList[intArrayList.Count - 1] + 23;
      }
      arrayListTimer.Stop();

      UnityEngine.Debug.Log(
        "Integer Array List time = " +
        (
(double)(arrayListTimer.Elapsed.TotalMilliseconds * 1000 * 1000) /
LENGTH
        ).ToString("0.00 nanoseconds average per cycle")
      );

      UnityEngine.Debug.Log(
        "Integer Array time = " +
        (
(double)(intArrayTimer.Elapsed.TotalMilliseconds * 1000 * 1000) /
LENGTH
        ).ToString("0.00 nanoseconds average per cycle")
      );
    }
  }
```

On Mac OS X, Intel Core i5 2.7 GHz, after testing this example in the Unity editor the following results were obtained:

```
Integer Array List time = 183.36 nanoseconds average per cycle
Integer Array time = 4.78 nanoseconds average per cycle
```

The difference is awesome and more than in the previous example, while using the class List. Thus, we demonstrated the obvious advantages of a simple one-dimensional array, compared with collections when large amounts of data can greatly ruin your performance. What cannot be said about the simple one-dimensional arrays is which of them are the building blocks for a variety of collections. As always, the choice is yours. The most important thing is not to forget the basic axioms in optimization decisions. Let's look at another example with the class Dictionary as shown in code example here:

```
using UnityEngine;
using System.Collections.Generic;
using System.Diagnostics;
```

```
public class DictionaryVsArray : MonoBehaviour {
  public const int CYCLES = 1000000;

  Dictionary<int, int> dictionary;
  int[] intArray;

  int tmpInt;

  void Start() {
    dictionary = new Dictionary<int, int>();
    intArray = new int[CYCLES];

    Stopwatch intArrayTimer = Stopwatch.StartNew();
    for (int i = 0; i < CYCLES; i++)
    {
      intArray[i] = i + 117;
      tmpInt = intArray[i] + 23;
    }
    intArrayTimer.Stop();

    Stopwatch dictionaryTimer = Stopwatch.StartNew();
    for (int i = 0; i < CYCLES; i++)
    {
      dictionary.Add(i, i + 117);
      tmpInt = (int)dictionary[dictionary.Count - 1] + 23;
    }
    dictionaryTimer.Stop();

    UnityEngine.Debug.Log(
      "Integer Dictionary time = " +
      (
(double)(dictionaryTimer.Elapsed.TotalMilliseconds * 1000 * 1000) /
CYCLES
      ).ToString("0.00 nanoseconds average per cycle")
    );

    UnityEngine.Debug.Log(
      "Integer Array time = " +
      (
(double)(intArrayTimer.Elapsed.TotalMilliseconds * 1000 * 1000) /
CYCLES
      ).ToString("0.00 nanoseconds average per cycle")
    );
  }
}
```

On Mac OS X, Intel Core i5 2.7 GHz, after testing this example in the Unity editor the following results were obtained:

```
Integer Dictionary time = 132.75 nanoseconds average per cycle
Integer Array time = 4.63 nanoseconds average per cycle
```

Also, I want to show you the following code example using the `Hashtable` collection for our performance testing, as shown in code example here:

```
using UnityEngine;
using System.Collections;
using System.Diagnostics;

public class HashtableVsArray : MonoBehaviour {
  public const int CYCLES = 1000000;

  Hashtable hashtable;
  int[] intArray;

  int tmpInt;

  void Start() {
    hashtable = new Hashtable();
    intArray = new int[CYCLES];

    Stopwatch intArrayTimer = Stopwatch.StartNew();
    for (int i = 0; i < CYCLES; i++)
    {
      intArray[i] = i + 117;
      tmpInt = intArray[i] + 23;
    }
    intArrayTimer.Stop();

    Stopwatch hashtableTimer = Stopwatch.StartNew();
    for (int i = 0; i < CYCLES; i++)
    {
      hashtable.Add(i, i + 117);
      tmpInt = (int)hashtable[hashtable.Count - 1] + 23;
    }
    hashtableTimer.Stop();

    UnityEngine.Debug.Log(
      "Integer Hashtable time = " +
      (
```

```
(double) (hashtableTimer.Elapsed.TotalMilliseconds * 1000 * 1000) /
CYCLES
        ).ToString("0.00 nanoseconds average per cycle")
    );

    UnityEngine.Debug.Log(
      "Integer Array time = " +
      (
(double) (intArrayTimer.Elapsed.TotalMilliseconds * 1000 * 1000) /
CYCLES
        ).ToString("0.00 nanoseconds average per cycle")
    );
  }
}
```

On Mac OS X, Intel Core i5 2.7 GHz, after testing this example in the Unity editor, the following results were obtained:

```
Integer Hashtable time = 539.59 nanoseconds average per cycle
Integer Array time = 4.52 nanoseconds average per cycle
```

As you can see, all the collections are significantly inferior to the simple one-dimensional array in performance, but the benefit in many situations is more convenient because of the use of a more readable code. However, you lose a lot of CPU time and memory, and have to sacrifice clarity of code. The rest of the collection and all other constructions you are interested in, you can easily perform your own tests for your system's performance, on the basis of the examples discussed earlier.

We will not leave without attention on loops that are used very often in code-like branching structure. Loops `for`, `while`, and `do-while` are the fastest compared to other cycles such as `foreach`. Another trick when using loops is that we can deploy loops for fewer passes as shown in code example here:

```
using UnityEngine;
using System.Diagnostics;

public class LoopsTest : MonoBehaviour {
  public const int CYCLES = 1000000;

  int [] tmpInt;
  int i, _optimizedCycles;

  void Start() {
    tmpInt = new int[CYCLES];
```

```
Stopwatch doWhileLoopTimer = Stopwatch.StartNew();
i = 0;
do
{
  // do while loop test
  tmpInt[i] = i + 123;
  i++;
} while (i < CYCLES);
doWhileLoopTimer.Stop();

Stopwatch whileLoopTimer = Stopwatch.StartNew();
i = 0;
while (i < CYCLES)
{
  // while loop test
  tmpInt[i] = i + 123;
  i++;
}
whileLoopTimer.Stop();

Stopwatch forLoopTimer = Stopwatch.StartNew();
for (i = 0; i < CYCLES; i++)
{
  // for loop test
  tmpInt[i] = i + 123;
}
forLoopTimer.Stop();

_optimizedCycles = Mathf.CeilToInt(CYCLES / 5);
Stopwatch optimizedTimer = Stopwatch.StartNew();
for (i = 0; i < _optimizedCycles; i++)
{
  // optimized for loop test
  tmpInt[i*5] = i*5 + 123;
  if (CYCLES > i*5+1) tmpInt[i*5+1] = i*5 + 124;
  if (CYCLES > i*5+2) tmpInt[i*5+2] = i*5 + 125;
  if (CYCLES > i*5+3) tmpInt[i*5+3] = i*5 + 126;
  if (CYCLES > i*5+4) tmpInt[i*5+4] = i*5 + 127;
}
optimizedTimer.Stop();

Stopwatch foreachTimer = Stopwatch.StartNew();
i = tmpInt.Length - 1;
foreach (int intElement in tmpInt)
```

```
      {
        // foreach test
        tmpInt[i] = intElement;
        i--;
      }
      foreachTimer.Stop();

    UnityEngine.Debug.Log(
      "Optimized For loop time = " +
      (
(double)(optimizedTimer.Elapsed.TotalMilliseconds * 1000 * 1000) /
CYCLES
      ).ToString("0.00 nanoseconds average per cycle")
    );

    UnityEngine.Debug.Log(
      "For loop time = " +
      (
(double)(forLoopTimer.Elapsed.TotalMilliseconds * 1000 * 1000) /
CYCLES
      ).ToString("0.00 nanoseconds average per cycle")
    );

    UnityEngine.Debug.Log(
      "While loop time = " +
      (
(double)(whileLoopTimer.Elapsed.TotalMilliseconds * 1000 * 1000) /
CYCLES
      ).ToString("0.00 nanoseconds average per cycle")
    );

    UnityEngine.Debug.Log(
      "Do While loop time = " +
      (
(double)(doWhileLoopTimer.Elapsed.TotalMilliseconds * 1000 * 1000) /
CYCLES
      ).ToString("0.00 nanoseconds average per cycle")
    );

    UnityEngine.Debug.Log(
      "Foreach time = " +
      (
(double)(foreachTimer.Elapsed.TotalMilliseconds * 1000 * 1000) /
CYCLES
```

```
        ).ToString("0.00 nanoseconds average per cycle")
    );
  }
}
```

On Mac OS X, Intel Core i5 2.7 GHz, after testing this example in the Unity editor the following results are obtained:

```
Optimized For loop time = 2.89 nanoseconds average per cycle
For loop time = 3.63 nanoseconds average per cycle
While loop time = 3.72 nanoseconds average per cycle
Do While loop time = 3.72 nanoseconds average per cycle
Foreach time = 5.62 nanoseconds average per cycle
```

The results speak for themselves. So, do not forget about optimizing your loops. The first step is to draw attention to the cycles with a huge number of passes, since they can lower your performance by several orders of magnitude. Before optimizing anything, you need to find the bottlenecks in your code and only then decide which optimization techniques you prefer or require.

As for the loop foreach, we recommend that you use this cycle only in exceptional situations. Let's look at a small sample code for each loop, and the next step as it is transformed into a completely different code loop. The next example shows a simple foreach loop:

```
foreach (YourType yt in yourCollection)
{
  yt.YourAction();
}
```

Next, let's look at what happens to a piece of code **foreach** loop. As we can see in the code example here, our cycle turns into a loop while using the enumerator object. The code is given here:

```
using (YourType.Enumerator e = this.yourCollection.GetEnumerator())
{
  while (e.MoveNext())
  {
    YourType yt = (YourType)e.Current;
    yt.YourAction();
  }
}
```

As for the characters, it is better to use a single character than a string consisting of a single character. A symbol is passed by value, and it is necessary and requires only two bytes of memory, while the string with one character requires more than 20 bytes of memory, since the string is passed by reference.

I would also like to mention the ToString function that is best used only when necessary, otherwise you risk degrading your performance. For example, using this function for characters is not always exact: there is rarely a decision justified. Generally, you should remember one simple and most important axiom of code optimization—with less code executed, less CPU time and memory is used that significantly improves your productivity. Let's look at a simple code example shown here, which covers one of the simplest options for optimizing the transfer of an integer to a string:

```
using UnityEngine;
using System.Diagnostics;

public class IntegerToStringTest : MonoBehaviour {
  public const int CYCLES = 1000;

  string str;

  void Start() {
    str = "";
    Stopwatch toStringTimer = Stopwatch.StartNew();
    for (int i = 0; i < CYCLES; i++)
    {
      str += i.ToString();
    }
    toStringTimer.Stop();

    str = "";
    Stopwatch optimizedToStringTimer = Stopwatch.StartNew();
    for (int i = 0; i < CYCLES; i++)
    {
      str += string.Empty + i;
    }
    optimizedToStringTimer.Stop();

    UnityEngine.Debug.Log(
      "ToString time = " +
      (
(double)(toStringTimer.Elapsed.TotalMilliseconds * 1000 * 1000) /
CYCLES
```

```
    ).ToString("0.00 nanoseconds average per cycle")
  );

  UnityEngine.Debug.Log(
    "Optimized ToString time = " +
    (
(double)(optimizedToStringTimer.Elapsed.TotalMilliseconds * 1000 *
1000) / CYCLES
    ).ToString("0.00 nanoseconds average per cycle")
  );
 }
}
```

On Mac OS X, Intel Core i5 2.7 GHz after testing this example in the Unity editor I got the following results:

```
ToString time = 18229.00 nanoseconds average per cycle
Optimized ToString time = 13576.00 nanoseconds average per cycle
```

Try to build your own tests for your functions on the basis of the preceding examples. In optimization, you should often trust your own instincts to find the cause of all ills, but to make the right decisions you need to rely on the results of the tests, that is, just pure mathematics and well-defined numbers.

Transforming Unity C# code into Unity JavaScript code and vice versa

The following is an example of how easy it is to convert your Unity C# code to Unity JavaScript code and vice versa. You can find on the web a lot of different automatic tools that you can use for free in order get conversation between Unity Scripts done as early as possible. As an example, you can convert Unity JS to Unity C# on this `http://www.m2h.nl/files/js_to_c.php`.

JavaScript variables and types

By default, the Unity JS code variables are public and visible in Unity Inspector. In order to hide your variables from Unity inspector or from other classes, you should mark all those variables with a `private` keyword as shown in the following simple examples:

```
// private variables are invisible in Unity Inspector
private var length : float = 2.9;
```

```
// visible in Unity Inspector
var title : String = "Title";

// visible in Unity Inspector
var isLoop : boolean = false;
```

C# variables and types

Float values in C# must have a lowercase **f** or an uppercase **F** at the end. Otherwise, it will be treated as a double value. Also notice that in JS code, the string type should be written with the first letter uppercase. However, in C# code with lowercase letters you can see our simple examples here:

```
// public variables are visible in Unity Inspector
public float length = 2.9f;

// is invisible in Unity Inspector
string title = "Title";

// is invisible in Unity Inspector
private bool isLoop = false;
```

By default, in Unity C# code, variables are private and invisible in the Unity inspector. In order to show your variables in the Unity inspector, you should mark all these variables with the `public` keyword as shown in our previous simple examples.

Converting types in Unity JS

You can use the following code to convert types in Unity JS:

```
var length : float = 0.08; // variable with float type
var number : int = length; // converting float to integer
print(number); // prints "0" in Unity console
```

Converting types in Unity C#

You can use the following code to convert types in Unity C#:

```
float length = 0.08f; // variable with float type
int number = (int)length; // converting float to integer
Debug.Log(number); // prints "0" in Unity console
```

Unity JS function versus Unity C# function

The following code snippet shows the syntax of the code as written in Unity JS and Unity C#

```
// Unity JS Function
function YourFunctionName (yourStringVarName : String) {
    print(yourStringVarName);
}

// Unity C# Function
public void YourFunctionName (string yourStringVarName) {
    Debug.Log(yourStringVarName);
}
```

Unity JS return versus Unity C# return

In Unity JS, you don't need to declare return type as shown in a simple function example here:

```
function JSReturnString() {
    return "Hello World !";
}
```

In Unity C#, you have to always declare a `return` type:

```
public string CSharpReturnString() {
    return "Hello World !";
}
```

Unity JS yielding versus Unity C# yielding

In JS yielding is quite simple, like the `return` keyword. You can just use yield statements without any declarations as shown in the simple example here:

```
function Start() {
    yield YourFunc();
    yield new WaitForSeconds(1.7);
    print("[Start] FINISH");
}

function YourFunc() {
    print("[YourFunc] START");
    yield new WaitForSeconds(0.8);
    print("[YourFunc] FINISH");
}
```

```
//Output will be similar as shown below:
// [YourFunc] START
// [YourFunc] FINISH
// [Start] FINISH
```

In C# code, you should declare the IEnumerator type in your method declaration as shown in the following example:

```
IEnumerator Start() {
    yield return StartCoroutine(YourMethod());
    yield return new WaitForSeconds(1.7f);
    Debug.Log("[Start] FINISH");
}

IEnumerator YourMethod() {
    Debug.Log("[YourMethod] START");
    yield return new WaitForSeconds(0.8f);
    Debug.Log("[YourMethod] FINISH");
}

//Output will be similar as shown below:
// [YourMethod] START
// [YourMethod] FINISH
// [Start] FINISH
```

Unity JS directives versus Unity C# directives

Unity has a number of **script directives**, for example, AddComponentMenu. The difference in syntax is shown in the following code:

```
// Unity JS example
@script AddComponentMenu ("Your Company Name/Your Action Name")
class YourFunctionName extends MonoBehaviour {}

// Unity C# example
[AddComponentMenu("Your Scope Name/Your Action Name")]
public class YourMethodName : MonoBehaviour {}
```

Summary

This chapter introduced different details about occlusion culling and LOD optimization techniques. Also, this chapter showed how to optimize Unity C# and Unity JS code. Finally, you saw main differences in syntax between Unity C# and Unity JS codes and learned how it is easy to transform between them.

In the next chapter, you will explore how to enhance the quality in games and applications using different techniques such as physically-based shaders and global illumination in Unity 5. At the end of the chapter, you will know how to optimize any shader code.

7
Troubleshooting and Best Practices

Primarily, this chapter will explore how to enhance the quality of games and applications using different techniques and physically-based shaders. Secondly, this chapter will describe global illumination in Unity 5. At the end of the chapter, you will optimize a shader code.

The topics that will be covered in the chapter are as follows:

- Measuring performance with the built-in Android profiler
- Debugging Android devices with the Unity profiler tool
- Best practices in scripts and shaders

Measuring performance with the built-in Android profiler

Let's look at what kind of information we can see from the built-in Android profiler in Unity 5.

General CPU activity

Next, we're going to talk about the information we can get from the built-in profiler. To make it easier to understand the structure of these messages, we'll cover them in groups. The first group includes the general information or in other words the overall performance statistics on the CPU.

You will see the total time that was spent by the CPU in the parameter value called `cpu-player`. The time that was spent on the CPU side by the OpenGLES driver code will be seen in the value of the parameter called `cpu-ogles-drv`. Next, let's consider the following parameter, known as `cpu-waits-gpu`. This option will not appear in the built-in profiler for very small values. This value shows how much CPU time was spent waiting for the end of rendering on the GPU side. Next, let's consider the following parameter, known as `msaa-resolve`. This value shows how much CPU time was spent on anti-aliasing methods. Let's consider the following parameter, known as `cpu-present`. This value shows how much CPU time was spent on executing the OpenGLES `presentRenderbuffer` function. Let's also look at the value of the last parameter in this group, known as `frametime`. This value shows the time spent on the CPU side for frame execution.

Refresh rate of the Android hardware is locked at about 60 Hz, so you will have the frame time at about ~16.7 ms (approximately 16.7 milliseconds we get from computing — 1000 milliseconds divided by 60 Hz).

Rendering statistics

Now let's consider the following group of statistics based on the rendering. This group contains only four parameters. The first parameter is called a `draw call`. The true meaning of this value is to show draw calls quantified per frame. The second parameter of this group is known as `tris`. This value indicates how many triangles the renderer will process. The third parameter of this group is known as `verts`. This value shows how many vertices the renderer will process. The upper limit number for static geometry is 10,000 vertices and much lower for skinned geometry. Finally, the last parameter, which we will study in this group, is called `batched`. The value of this parameter greatly affects your performance, so try to reduce this value as much as possible. This value demonstrates the number of automatically batched draw calls, triangles, and vertices by the Unity engine.

In order to improve Unity engine batching, you should use shared materials everywhere possible for all available objects.

Detailed Unity player statistics

Now, consider the following group of statistics, which is more detailed. The first parameter in the detailed statistics of the built-in profiler is known as physx. This value indicates the time spent on physics engine execution. The text parameter is called animation. This value indicates the time spent on bone animations. The third parameter in the detailed statistics is called culling. This value indicates how much time was spent culling the object's execution. The fourth parameter in the detailed statistics of the built-in profiler is known as skinning. This value indicates the time we need to apply animations to skinned meshes. The fifth parameter in this detailed statistics is called batching. This value shows the time spent on batching geometry execution.

 Batching static geometry is less expensive versus batching dynamic geometry.

The sixth parameter in the detailed statistics is called render. This value represents the execution time spent on rendering visible objects. The seventh parameter is called fixed-update-count. This value shows the upper and lower values of the FixedUpdate execution time for the current frame. Try to decrease this value as much as possible because it can decrease your performance.

Detailed script statistics

There are just three obtainable parameters. The first one is known as update. This value determines the time used for execution per Update function in your scripts. The next parameter is called fixedUpdate. This value demonstrates the time used for all executions per FixedUpdate function in your scripts. The following parameter is known as coroutines. This value determines the time utilized for coroutines execution in your scripts.

Detailed statistics on memory allocated by scripts

Let's cover the following group of statistics based on detailed statistics for memory allocation by your scripts. There are only four parameters. The first parameter is called allocated heap.

This value represents available memory for allocation. If we need more memory than is available in the heap, a garbage collector will be called. However, if the garbage collector cannot free up more memory for us, then the heap will be increased in size. The next parameter is known as `used heap`. This value indicates the allocated heap by objects. It will be increased for each new class instance, and not for structs before the garbage collector will be called one more time. The following parameter is known as `max numbers of collections`. This value shows the quantity of the garbage collector calls within the last 30 frames. The last parameter in this group, and the last one in the built-in profiler, is called `collection total duration`. This value displays the summarized milliseconds used for the garbage collector calls within the last 30 frames.

Debugging Android devices with the Unity profiler tool

We can open the Unity profiler window from the menu, which presents the whole Unity profiler tool. In the upcoming sections, we will explore more about the Unity profiler areas.

Before starting, we need to know how this tool is works and how simple it is to use. Firstly, let's look more at the Unity profiler tool window structure and separate its parts. As we can see in the next screenshot, there are four main visual parts:

- Profiler controls
- Usage area
- Profiler timeline
- Information table

The upcoming sections focus on these distinctive parts of the Unity profiler tool. Let's dive into the most interesting thing in this instrument.

Regarding the visual profiler, you can connect to access devices on which your application is performing in order to further analyze the performance of your software. In order to connect to the other device, it is necessary (but not just sufficient) for the profiler to be on the same local network. The **Active Profiler** option allows you to select your device from a list of the desired connections. Besides that, your application should be launched with the **Development Build enabled** checkbox from **Build Settings**. Also in these settings, you will see the **Autoconnect profiler** option, which is necessary to signal whether Unity should or should not be connected to the profiler every time you start your application.

The following are the Unity profiler buttons:

- **Record**
- **Deep Profiler**
- **Profile Editor**
- **Active Profiler**
- **Clear**

If you look at the top of the profiler window, you will see there is a toolbar, which we will examine in more detail later in the chapter. Using the buttons on the toolbar, you can enable or disable the profiler recording data. Also, you can clear the collected information or navigate in the frame set and much more; we will talk more about this later. Right in the toolbar, we see the **Current** button. After clicking on this button, we will automatically get on a frame and the last detail of its implementation. If your game was played in the Unity editor, it will be suspended, meaning it will be paused. It will also be suspended when switching frames forward or backward, using the arrows buttons, which are not far away from the **Current** button. Also, be aware that the profiler does not preserve all the frames, but only a certain number of the most recent frames. Furthermore, if you go from left to right on the toolbar profiler, we see a **Clear** button to clear all the data that was collected. After this, we see a **Active Profiler** button, which allows you to select a device or the Unity editor for further performance analysis.

Next, we see a button called **Profile Editor**; if you click on this button, you will begin to get detailed statistics execution for the Unity editor. To the left of this button, you will see the **Deep Profile** button. When this button is activated, it will provide information about all of your scripts and function calls. Deep profiling can significantly slow down your application or your game, as it will be necessary to spend most of the time processing and requires a huge amount of memory space. Remember that very deep profiling will only work if you use it for small projects, otherwise you run the risk that Unity will not be able to obtain the necessary resources and hang, following which you will have to restart the Unity editor. Also, deep profiling is well suited not only for small projects, but is also very useful for testing key aspects of your game or application. You can use the code in deep profiling, and that is switched ON and OFF for specific pieces of code in your scripts. Only the necessary parts of the code will be profiled and analyzed.

The `Profiler.BeginSample` and `Profiler.EndSample` calls are the beginning and endpoints, respectively, of profiling your code, which means that the code between these two function calls will be profiled and detailed statistics will be displayed in the bottom profiler window. We will talk about profiler scripting a little later in this chapter. On the left from the **Deep Profile** button is a button that is called **Record**, and it is needed to enable or disable profiling as we mentioned earlier. Well, the left-most button in the toolbar is called **Add Profiler,** and it is needed to display different profiler areas: **CPU, GPU, Rendering, Memory, Audio, Physics 3D,** and **Physics2D.** We will talk about these profiler areas later in this chapter.

If your game or your application is running at a specific frame rate or is synchronized with the vertical blank, then Unity will keep the average time waiting for the synchronization of all frames in a parameter called `Wait For Target FPS`, which is displayed in the profiler. By default, the information on waiting times will not be published in the Unity profiler. To change the specified default behavior, you need to enable `View SyncTime`.

Profiler timeline

In the upper part of the profiler window is a graph that shows the profiler load data in real time. Statistics are processed in each frame and are saved only in the history of the last couple of hundred frames. If you select one of the frames for further consideration, you will see details at the bottom profiler, which in turn will depend on the selected timeline area (for example, **CPU, GPU,** or **Audio**). You can both add and remove various timeline areas. Also, note that the colored squares on the left display different timeline areas. In fact, it is not just the colored squares; they are radio buttons. Thus, it will be much easier to eliminate unnecessary data when optimizing your application.

The CPU area

The CPU area clearly shows which specific place and how much total time was spent on the CPU side of course. If you choose it, then you will have hit on the CPU area. After that, you will see that the bottom profiler displays enough details solely about the execution of your application on the CPU. Also, you can choose two different modes of displaying detailed information:

* The **Hierarchy** mode displays the information in a hierarchy, by grouping the data
* The **Group Hierarchy** mode displays information on groups that were distributed logically; for example, the **Rendering** group, the **Scripts** group, the **Physics** group, and many more groups

The **Others** area of the CPU profiler incorporates **Loading**, **Audio**, **Animation**, **Particles**, **Playerloop**, **AI**, and **Networking**.

The Rendering area

The **Rendering** area displays rendering statistics as shown in the following screenshot. The timeline graphically presents the number of rendered **Draw Calls**, **Triangles**, and **Vertices**. As we can see, the bottom part of the following screenshot and **Game View Rendering Statistics** shown in the next figure are very similar. Further, we will cover the information shown in the following screenshot in details:

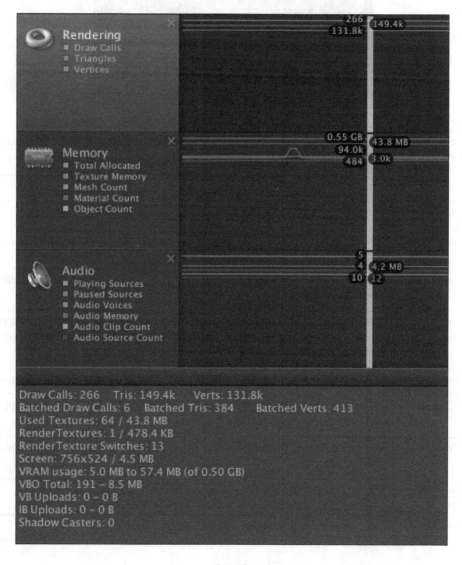

The following screenshot is very similar to the same statistics information:

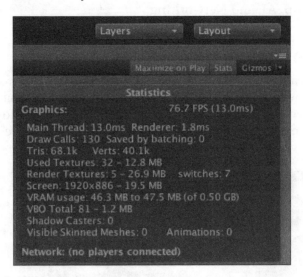

Time per frame and FPS	The time spent rendering one frame in milliseconds; represents quantity of frames per second.
Draw calls	The amount of rendered meshes.
Batched (draw Calls)	The number of batched draw calls.
Tris and verts	The amount of drawn geometry (triangles and vertices).
Used textures	This indicates how many textures were used and the amount of memory needed per frame.
Render textures	This shows the amount of times active render texture was switched per frame; furthermore it demonstrates how much memory is used to render texture, and how many render textures there were.
Screen	This shows the screen size with its anti-aliasing level and memory usage.
VRAM usage	This roughly indicates the amount of video memory (VRAM) usage; furthermore, represents how much memory your graphic card has.
VBO total	**Vertex Buffers Objects (VBO)** is the number of uploaded meshes to the graphics card.
Visible Skinned Meshes	This shows the amount of rendered skinned meshes.
Animations	This represents how many animations can be played.

The Memory area

While profiling this area, you can choose one of the two available modes for different displaying modes. The first mode is to display very simple statistics, and the second mode is for displaying very detailed statistics. We cover these two modes in more detail in the following sections:

The simple view

We begin with the simplest display mode statistics (as shown in the following screenshot). This shows the usage of memory for each profiled frame in a simpler form than it does in the detailed statistics:

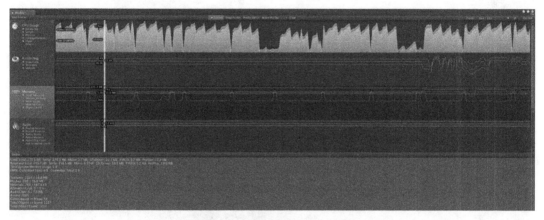

The simple view of the Unity profiler

In order to use memory efficiently, Unity tries to keep a certain amount of memory in advance in the form of a pool, or in other words, as a backup buffer, which greatly improves performance. Statistics memory, or rather information about how much memory is consumed, and for what is the described method of reservation, will be shown at the bottom profiler window. Here are the parameters of these statistics:

- **Unity**: This indicates the amount of memory used for allocations in the native Unity code
- **Mono**: This shows how big the heap size was, and the amount of memory used for the garbage collector
- **Gfx Driver**: This indicates the amount of memory used by the driver on shaders, meshes, render targets, and textures
- **FMOD**: This shows the amount of memory used on audio drivers
- **Profiler**: This indicates the amount of memory used for the Unity's profiler

The memory area displays information for fundamental types of objects and assets: textures, meshes, materials, animations, audio, and object count.

The detailed view

In a detailed view, you can save the current state for further analysis using the **Take Sample** button. In order to obtain such detailed information about the memory usage, the Unity profiler should take time to collect all information needed, and that's why you should not think that you can receive information in real time.

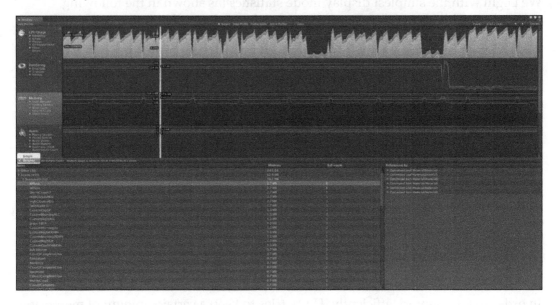

The profiler will show information about where and on what the memory was consumed. The following is a list of groups of objects that will be spent in memory:

- Referenced from native code
- Scene object
- Built-in resources
- Marked as don't save

After you click on one of the objects in the list, Unity will highlight the selected items in the **Project** view or in the **Scene** view. When profiling your application in the Unity editor, the statistics will be less accurate than it could be on a particular device. Some of the costs associated with the Unity editor execution will also be displayed in the average values that will not be true for your application. Therefore, for a more precise analysis of your application, it is the best decision to connect to real devices and profile statistics in that case.

The audio area

This shows the information displayed in the audio area.

The physics area

The following is a list of information displayed in the Physics 3D area (as shown in following screenshot):

- **Active Bodies**: This indicates the number of awake Rigidbodies
- **Sleeping Bodies**: This displays the number of sleeping Rigidbodies
- **Number of Contacts**: This shows the total amount of contact points in the scene between all colliders
- **Static Colliders**: This represents how many colliders were attached on non-Rigidbody objects
- **Dynamic Colliders**: This demonstrates how many colliders were attached on Rigidbody objects

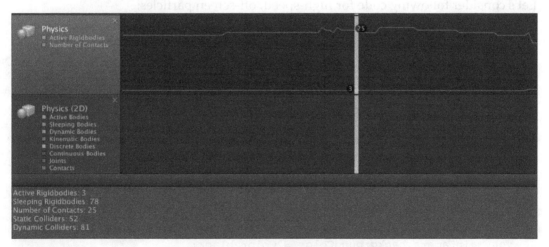

The detailed view of the Unity profiler

The GPU area

Statistics that are displayed in the profiler window for the GPU area are very similar to the displayed statistics for CPU area.

 On the Mac, only OSX 10.7 Lion and later versions support GPU profiling.

Real-practice techniques

There are two different performance optimization techniques that are used by many professional developers from all over the world.

The high-speed, off-screen particles technique in Unity

The next technique is to optimize the particle system, which was introduced by NVIDIA, GPU Gems 3. The first step in order to achieve the goal is to render particles into RenderTexture or, in other words, into another render target with smaller size than screen. The second step in this idea is to blend the particles back into screen. First, we need depth buffer. When we render into another render target, we need depth buffer for its z-testing. In the following line of code, you can register in the Awake or Start callbacks just as examples:

```
this.camera.depthTextureMode = DepthTextureMode.depth;
```

Let's consider following code for high-speed, off-screen particles:

```
// create the off-screen particles texture
RenderTexture yourParticlesRenderTexture = RenderTexture.GetTemporary(
  Screen.width, // yourLowerResolutionIntegerValue
  Screen.height, // yourLowerResolutionIntegerValue
  0
);
```

The yourLowerResolutionIntegerValue determines the quality. The highest value means the worst quality and the best performance and vice versa.

The second part is very simple and means just tuning your main camera's properties as shown here:

```
yourMainCamera.targetTexture = yourParticlesRenderTexture;
yourMainCamera.backgroundColor = Color.black;
yourMainCamera.cullingMask = yourLayerMask.value;
yourMainCamera.depthTextureMode = DepthTextureMode.None;
yourMainCamera.clearFlags = CameraClearFlags.SolidColor;
```

The next step includes rendering and blending particles into the scene:

```
Shader.SetGlobalVector(
  "_Your_Camera_Depth_Texture_Size",
  Vector4(
```

```
      this.camera.pixelWidth, this.camera.pixelHeight, 0.0, 0.0
   )
);

depthCamera.RenderWithShader(
   Shader.Find("Pro/Unity/Performance/Particles/Off-Screen"),
   "RenderType"
);
Material yourMixedMaterial = YouClassHelper.GetMaterialByShader(
   Shader.Find("Pro/Unity/Performance/Particles/Off-Screen")
);

Vector2 yourTexelOffset = Vector2.Scale(
   source.GetTexelOffset(),
   Vector2(source.width, source.height)
);

Graphics.BlitMultiTap(
   yourParticlesRenderTexture, source, yourMixedMaterial,
yourTexelOffset
);
```

 Always release the particles render texture for better performance.

You can render (after postprocessing) your `RenderTexture` to your destination as shown here:

```
RenderTexture.ReleaseTemporary(yourParticlesRenderTexture);
Graphics.Blit(source, destination);
```

The pool technique

The next technique is a basic pooling system (as shown in Listing 3-1) for Unity in addition for Shuriken particles. Put the pool component on your GameObject and set the name and prefab. The pool summons the `OnCreateEvent` strategy on entities when they are *made* in the pool (so put your initialization that typically will go in the `Start` or `Awake` callback) and an `OnLiberationEvent` system when reused items go into the pool. The `OnCreateEvent` strategy provides the pool that made the occurrence so that you can store it away and reuse your **GameObject** later:

```
YourPoolClass.cs
using UnityEngine;
using System.Collections;
using System.Collections.Generic;
```

```
public class YourPoolClass : MonoBehaviour
{
    private static readonly Dictionary<string, YourPoolClass>
    namesOfObjects = new Dictionary<string, YourPoolClass>();

    public static YourPoolClass GetPoolByName(string name) {
    return namesOfObjects[name];
  }

    [SerializeField]
    private string nameOfYourPool = string.Empty;

    [SerializeField]
    private Transform yourPoolPrefab = null;

    [SerializeField]
    private int initialObjectCounter = 23;

    [SerializeField]
    private bool isParentEnabled = true;

    private readonly Stack<Transform> yourObjectsStack = new
Stack<Transform>();

    void Awake()
    {
    System.Diagnostics.Debug.Assert(yourPoolPrefab);
    namesOfObjects[nameOfYourPool] = this;

        for (int i = 0; i < initialObjectCounter; i++)
        {
            var t = Instantiate(yourPoolPrefab) as Transform;
            AdjustingYourObject(t);
            LiberationObject(t);
        }
    }

    public Transform GetObject(Vector3 position = new Vector3())
    {
        Transform t = null;

        if (yourObjectsStack.Count > 0)
    {
            t = yourObjectsStack.Pop();
```

```
        }
    else
    {
            Debug.LogWarning(
        nameOfYourPool + " pool error!", this
    );
            t = Instantiate(yourPoolPrefab) as Transform;
        }

        t.position = position;
        AdjustingYourObject(t);

        return t;
    }

    private void AdjustingYourObject(Transform obj)
    {
        if (isParentEnabled)
        {
            obj.parent = transform;
        }

        obj.gameObject.SetActiveRecursively(true);
        obj.BroadcastMessage(
    "OnCreateEvent",
    this,
    SendMessageOptions.DontRequireReceiver
);
    }

    public void LiberationObject(Transform obj)
    {
        obj.BroadcastMessage(
    "OnLiberationEvent",
    this,
    SendMessageOptions.DontRequireReceiver
);
        obj.gameObject.SetActiveRecursively(false);
        yourObjectsStack.Push(obj);
    }
}
```

This is how to use this pooling system:

```
using UnityEngine;

public class YourPoolExampleUsage : MonoBehaviour {
  void Start() {
    YourPoolClass pool = YourPoolClass.GetPoolByName("Bang");
    Transform obj = pool.GetObject(Vector3.zero);
  }
}
```

In event of using a particle system with YourPoolClass, you should use the following code:

```
using UnityEngine;
using System.Collections;

[RequireComponent(typeof(ParticleSystem))]
public class YourPoolParticleSystem : MonoBehaviour
{
    private YourPoolClass yourPoolClass;

    void OnCreateEvent(YourPoolClass ypc)
    {
        yourPoolClass = ypc;

        particleSystem.renderer.enabled = true;
        particleSystem.time = 0;
        particleSystem.Clear(true);
        particleSystem.Play(true);
    }

    void OnLiberationEvent()
    {
        particleSystem.Stop();
        particleSystem.time = 0;
        particleSystem.Clear(true);
        particleSystem.renderer.enabled = false;
    }

    void Update()
    {
        if (!particleSystem.IsAlive(true) && particleSystem.renderer.
enabled)
```

```
    {
            yourPoolClass.LiberationObject(transform);
    }
  }
}
```

The scriptable profiler tool

The fact that developers can use the Unity profiler for profiling their own code or certain pieces of code is very important. In order to display statistics information about some of your function or for some part of your code in the Unity profiler, you just need to include your code between two calls, `Profiler.BeginSample` and `Profiler.EndSample`. After that you can use the visual Unity profiler tool to search for bottlenecks and spikes in your code.

 The profiler is only available in Unity Pro. In standalone games, the profiler can dump all profiling information using `Profiler.log` and `Profiler.enabled`.

To create your own tool, you can utilize the following Unity API calls:

- `FindObjectsOfTypeAll`
- `FindObjectsOfType`
- `GetRuntimeMemorySize`
- `GetMonoHeapSize`
- `GetMonoUsedSize`
- `Profiler.BeginSample`
- `Profiler.EndSample`
- `UnloadUnusedAssets`
- `System.GC.GetTotalMemory`
- `Profiler.usedHeapSize`

Unity profiler tricks

There is the capability to export the profiling information to a binary file, which can then be imported again later. This is empowered through scripting by means of:

```
function Start () {
        Profiler.logFile = "yourName.log";
```

```
            Profiler.enableBinaryLog = true; // writes to "yourName.log.
    data"

            Profiler.enabled = true;
    }
```

And reimported into the profiler:

```
    Profiler.AddFramesFromFile ("yourName.log");
```

The API to get to the profiler frame information into the script is uncovered in:

```
    UnityEditorInternal.ProfilerDriver
```

Not documented, yet totally open in the `UnityEditorInternal` namespace. Other suitable APIs:

```
    Profiler.BeginSample("Your Label Name");
    Profiler.EndSample();
    Profiler.GetRuntimeMemorySize(o : Object) : int
```

Creating a simple profiler

Now is the time to develop our own simple and very useful profiler tool from scratch. In the future, you will be able to use these scripts from our simple profiler for all of your projects as well as any other examples discussed in this book. Of course, you can modify all methods to meet your specific problem if you have a strong desire or if you have to do it, or you can use them all in their original form if this functionality will be enough for your tasks. First, let's look at a very simple class, which is a core class in our simple code profiler tool. In the following code you can see a very simple `ExampleProfilerClass`:

Listing 1-3. ExampleProfilerClass.cs

```
    using UnityEngine;

    public class ExampleProfilerClass
    {
      int counter = 0;

      float startedTime = 0;
      float totalTime = 0;
      float endTime = 0;
      float elapsedTime = 0;

      bool wasStartedFlag = false;
```

```
    public string indexStr;

    public float TotalTime {
      get {
        return totalTime;
      }
    }

    public int Counter {
      get {
        return counter;
      }
    }

    public ExampleProfilerClass(string indexStr)
    {
      this.indexStr = indexStr;
    }

    void ShowError() {
      Debug.LogError("ExampleProfilerClass {START / END} ERROR: [index]
= [" + indexStr + "]");
    }

    public void Start() {
      if (wasStartedFlag) {
        ShowError();
      }

      counter++;

      wasStartedFlag = true;

      startedTime = Time.realtimeSinceStartup;
    }

    public void End() {
      endTime = Time.realtimeSinceStartup;

      if (false == wasStartedFlag) {
        ShowError();
      }

      wasStartedFlag = false;
```

```
      elapsedTime = (endTime - startedTime);

      totalTime += elapsedTime;
   }

   public void ClearStatistics() {
      wasStartedFlag = false;

      totalTime = 0;

      counter = 0;
   }
}
```

You need to attach the `ExampleProfilerClass` script to one of the objects in your scene. The code is very straightforward and simple, as are all the other examples in this book. The entire code of our profiler is shown in Listing 1-4:

Listing 1-4. SimpleProfiler.cs

```
using UnityEngine;
using System.Collections.Generic;

public class SimpleProfiler : MonoBehaviour {
   float startedTime = 0;
   float followingTime = 1;
   float totalTimeInMilliSeconds = 0;
   float averageTimeInMilliSeconds = 0;
   float framesPerSecond = 0;
   float savedTimeInMilliSeconds = 0;
   float percentageSavedFromTotal = 0;
   float timeInMilliSecondsPerFrame = 0;
   float timeInMilliSecondsPerCall = 0;
   float callsNumberPerFrame = 0;

   int frameCount = 0;
   int colWidth = 30;

   static Dictionary<string, ExampleProfilerClass> statistics = new
Dictionary<string, ExampleProfilerClass>();

   string profilerInfo = "ALREADY STARTED !";

   Rect windowRect = new Rect(25, 25, 800, 300);
```

```
void Awake() {
  startedTime = Time.time;
}

void OnGUI() {
  GUI.Box(windowRect,"Simple Profiler");
  GUI.Label(windowRect, profilerInfo);
}

public static void Start(string indexStr) {
  if (false == statistics.ContainsKey(indexStr)) {
    statistics[indexStr] = new ExampleProfilerClass(indexStr);
  }

  statistics[indexStr].Start();
}

public static void End(string indexStr) {
  statistics[indexStr].End();
}

void Update() {
  frameCount++;

  if (Time.time > followingTime)
  {
    profilerInfo = "\n\n\n";

    totalTimeInMilliSeconds = (Time.time - startedTime) * 1000;
    averageTimeInMilliSeconds = (totalTimeInMilliSeconds /
frameCount);
    framesPerSecond = (1000 / (totalTimeInMilliSeconds /
frameCount));

    profilerInfo += "Frames per Second: ";
    profilerInfo += framesPerSecond.ToString("0.#") + " frames; \
nAverage Frame Time: ";
    profilerInfo += averageTimeInMilliSeconds.ToString("0.#") + " ms
\n\n\n";
    profilerInfo += "Time Percentages".PadRight(colWidth);
    profilerInfo += "ms per Frame".PadRight(colWidth);
    profilerInfo += "ms per Call".PadRight(colWidth);
    profilerInfo += "Calls number per Frame".PadRight(colWidth);
```

```
        profilerInfo += "NameIndex";
        profilerInfo += "\n";

        foreach(ExampleProfilerClass statisticsRecord in statistics.
Values)
        {
            savedTimeInMilliSeconds = (statisticsRecord.TotalTime * 1000);
            percentageSavedFromTotal = (savedTimeInMilliSeconds * 100) /
totalTimeInMilliSeconds;
            callsNumberPerFrame = statisticsRecord.Counter / (float)
frameCount;
            timeInMilliSecondsPerCall = savedTimeInMilliSeconds /
statisticsRecord.Counter;
            timeInMilliSecondsPerFrame = savedTimeInMilliSeconds /
frameCount;

            profilerInfo += (percentageSavedFromTotal.ToString("0.000") +
"%").PadRight(colWidth);
            profilerInfo += (timeInMilliSecondsPerFrame.ToString("0.000")
+ " ms").PadRight(colWidth);
            profilerInfo += (timeInMilliSecondsPerCall.ToString("0.0000")
+ " ms").PadRight(colWidth);
            profilerInfo += (callsNumberPerFrame.ToString("0.000")).
PadRight(colWidth);
            profilerInfo += (statisticsRecord.indexStr);
            profilerInfo += "\n";

            statisticsRecord.ClearStatistics();
        }

        frameCount = 0;

        startedTime = Time.time;

        followingTime = Time.time + 1;
    }
  }
}
```

The following is a simple test code, which performs mathematical operations in a cycle. You can hang this script on any object (or on multiple objects simultaneously) in your scene, just for testing your `SimpleCodeProfiler` tool.

Listing 1-5. TestProfilerCode.cs

```csharp
using UnityEngine;

public class TestProfilerCode : MonoBehaviour {
  float tmpFloat;

  void Update () {
    SimpleProfiler.Start("YOUR_UNIQUE_LABEL_1");

    for (int i = 0; i < 10; i++) {
      for (int degree = 0; degree < 360; degree++) {
        tmpFloat = Mathf.Cos(degree * Mathf.Deg2Rad);
      }
    }

    SimpleProfiler.End("YOUR_UNIQUE_LABEL_1");

    //////////////////////////////////////////////////

    SimpleProfiler.Start("YOUR_UNIQUE_LABEL_2");

    for (int i = 0; i < 50; i++) {
      for (int degree = 0; degree < 180; degree++) {
        tmpFloat = Mathf.Sqrt(Mathf.Cos(degree * Mathf.Deg2Rad) +
Mathf.Sin(degree * Mathf.Deg2Rad));
      }
    }

    SimpleProfiler.End("YOUR_UNIQUE_LABEL_2");
  }
}
```

Summary

In this chapter, we researched our choices for optimization in Unity. We first discovered different Unity performance areas. We explored the in-built Unity profiler and it's log information structure. In this chapter, we particularly discussed the Unity's profiler tool and its window parts. We discovered how to attach the profiler to different platforms and devices. At the end of this chapter, we talked about best practices that are used by many professionals. We also discovered the Unity profiler programming area and created our own very simple profiler tool.

In the bonus chapter, which is available online, I will show you how easy it is to develop the most popular game on Android Play Store (Glow Hockey has about 100,000,000–500,000,000 downloads at `https://play.google.com/store/apps/details?id=com.natenai.glowhockey&hl=en`) in Unity 5 from scratch. You will see how to create a camera for any screen resolutions and any screen sizes. Also, there you will see, in practice, how easy it is to use physics. You will learn in practice how to design beautiful effects, animations, physical behaviors, and other different real-world features and techniques for your Android games and applications. You will see how to optimize your project and any other real-world projects for Android devices. Many more useful things and features will be covered in the chapter.

Index

A

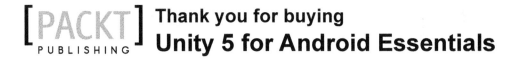

About Packt Publishing

Packt, pronounced 'packed', published its first book, *Mastering phpMyAdmin for Effective MySQL Management*, in April 2004, and subsequently continued to specialize in publishing highly focused books on specific technologies and solutions.

Our books and publications share the experiences of your fellow IT professionals in adapting and customizing today's systems, applications, and frameworks. Our solution-based books give you the knowledge and power to customize the software and technologies you're using to get the job done. Packt books are more specific and less general than the IT books you have seen in the past. Our unique business model allows us to bring you more focused information, giving you more of what you need to know, and less of what you don't.

Packt is a modern yet unique publishing company that focuses on producing quality, cutting-edge books for communities of developers, administrators, and newbies alike. For more information, please visit our website at www.packtpub.com.

Writing for Packt

We welcome all inquiries from people who are interested in authoring. Book proposals should be sent to author@packtpub.com. If your book idea is still at an early stage and you would like to discuss it first before writing a formal book proposal, then please contact us; one of our commissioning editors will get in touch with you.

We're not just looking for published authors; if you have strong technical skills but no writing experience, our experienced editors can help you develop a writing career, or simply get some additional reward for your expertise.

Learning Unity Android Game Development

ISBN: 978-1-78439-469-1   Paperback: 338 pages

Learn to create stunning Android games using Unity

1. Leverage the new features of Unity 5 for the Android mobile market with hands-on projects and real-world examples.

2. Create comprehensive and robust games using various customizations and additions available in Unity such as camera, lighting, and sound effects.

3. Precise instructions to use Unity to create an Android-based mobile game.

Mastering Unity Scripting

ISBN: 978-1-78439-065-5   Paperback: 380 pages

Learn advanced C# tips and techniques to make professional-grade games with Unity

1. Packed with hands-on tasks and real-world scenarios that will help you apply C# concepts.

2. Learn how to work with event-driven programming, regular expressions, customized rendering, AI, and lots more.

3. Easy-to-follow structure and language, which will help you understand advanced ideas.

Please check **www.PacktPub.com** for information on our titles

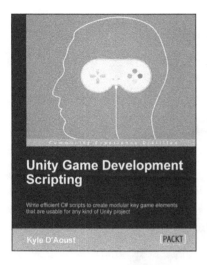

Unity Game Development Scripting

ISBN: 978-1-78355-363-1   Paperback: 202 pages

Write efficient C# scripts to create modular key game elements that are usable for any kind of Unity project

1. Write customizable scripts that are easy to adjust to suit the needs of different projects.

2. Combine your knowledge of modular scripting elements to build a complete game.

3. Build key game features, from player inventories to friendly and enemy artificial intelligence.

Learning Unreal® Engine iOS Game Development

ISBN: 978-1-78439-771-5   Paperback: 212 pages

Create exciting iOS games with the power of the new Unreal® Engine 4 subsystems

1. Learn about the entire iOS pipeline, from game creation to game submission.

2. Develop exciting iOS games with the Unreal Engine 4.x toolset.

3. Step-by-step tutorials to build optimized iOS games.

Please check **www.PacktPub.com** for information on our titles